Getting Horses Fit

JULIET

Getting Horses Fit
A Guide to Improving Performance

Sarah Pilliner

COLLINS
8 Grafton Street, London W1

Collins Professional and Technical Books
William Collins Sons & Co. Ltd
8 Grafton Street, London W1X 3LA

First published in Great Britain by
Collins Professional and Technical Books 1986

Distributed in the United States of America
by Sheridan House, Inc.

British Library Cataloguing in Publication Data
Pilliner, Sarah
Getting horses fit: a guide to improving performance.
1. Horses—Training
I. Title
636.1'083   SF287

ISBN 0–00–383197–3

Typeset by V & M Graphics Ltd, Aylesbury, Bucks
Printed and bound in Great Britain by
Mackays of Chatham, Kent

Contents

Foreword

With the dramatic increase in equestrian activities worldwide and with so many new ideas being tried out alongside the traditional methods of horse management, an authoritative guide, such as this one, which explains exactly what is required by today's equines to meet modern demands, is essential.

Performance is the essential element to success in whatever sphere but, without the knowledge of how this can and should be achieved, comparatively few horses reach their true potential because of lack of correct care and preparation for the job they have been set.

Getting Horses Fit – A guide to improving performance, seems to fill the gap authoritatively and sensibly on all aspects of stable management and fitness training. It is an excellent source of reference for the new horse owner as well as for the experienced horseman or trainer, explaining extremely effectively in a simple but detailed style not only how but why each aspect of horse care is so important. It also gives guidance on training programmes for the different types of competitions and explains how to assess the degree of fitness and condition.

Sarah Pilliner's book will certainly be a welcome addition to the bookshelf of every caring horseman and its contents will contribute greatly to the welfare of the horse.

Jane Holderness-Roddam

Preface

To enable a competition horse to give its best performance it is essential to provide the horse with a combination of correct feeding and exercise. The aim of this book is to explain in simple terms the changes that take place in the horse's body during exercise and training; this knowledge can then be used to devise suitable training programmes for disciplines as diverse as long-distance riding and three-day eventing. It is hoped that an insight into the inner workings of the horse will be of equal interest to trainers, riders and owners, be they preparing horses for Badminton, hunting, racing or Pony Club events.

Horse-men should be aware of the consequences of overstressing their horses, and in particular they should be concious of the stresses imposed by overcompeting, competing when unfit and travelling long distances. The horseman who has an understanding of basic exercise physiology, i.e. how the horse's body works, will be able to use the knowledge to benefit both himself and his horses in terms of better performance, improved welfare of the horse and better economics. If a horseman can design better training programmes the result will be a fitter horse and, since unfitness contributes to problems such as tendon breakdown, fitness will produce a longer-lasting horse. The fitter horse will compete not only more often but also more successfully.

To train a horse for any sport is a time-consuming and expensive business. This book will enable that time and money to be spent more effectively.

Acknowledgments

Thanks for their patience and time are due to Jeremy Houghton Brown, Head of the Horse Department, Warwickshire College of Agriculture, and David Jagger BVSc, MRCVS for reading and commenting on the initial draught. My thanks also to Mrs Heather Hathaway for the hours spend deciphering my writing and typing. All photographs are by Alison Francis unless otherwise stated.

Part I
Achieving Fitness

1 The Aim of Training

Fitness can be defined as being in a state suitable for an effort or act, and the level of fitness will be determined by the type of effort or act that is to be performed. In other words, each equestrian sport will demand a specific level and type of fitness from the horse, and a horse fit for one sport may not be fit for another. A showjumper capable of jumping Grade A courses with no distress would not be fit enough to complete a three-day event without being distressed.

To obtain fitness, to get a horse fit or to condition it, a combination of correct work and feeding is necessary. Together, work and feeding are known as 'training'. It is important at this stage to establish what is meant by work: a physics textbook will tell you that work is 'the effort made when a load is moved over a set distance'. This is easily understood in terms of a horse carrying a rider or pulling a load over a set course.

The aim of getting a horse fit, or 'conditioning' a horse, is to enable the horse to undergo a set amount of work with mimumum fatigue. Obviously, the degree of fitness necessary will depend on the severity of the work. The basis of any method of conditioning is to increase the horse's ability to tolerate work by giving it gradually increasing amounts of work to do during its training programme. This slow, steady progression of exercise levels is fundamental for both the physical and mental well-being of the horse, building confidence and security as a result of individual attention.

'Winning condition' is built and maintained on a foundation of good health. Athletic performance makes great demands on the horse's body; a healthy body is better able to withstand the stress of physical activity. The trainer and rider must be aware of the signs of good health and be able to respond to the most subtle indications of ill health.

The three most easily monitored signs of health are temperature, pulse and respiration. The values of these vital signs should fall within a 'normal' range. However, each horse will vary slightly and the trainer should know the values for each individual horse. Unless the horse's normal temperature, pulse and respiration (T.P.R.) are known, any deviation from normal, indicating stress, will not be easy to evaluate. The T.P.R. should be measured when the horse is calm and at rest (the method is described in detail in chapter 15).

The horse's normal temperature range is about 38°C (100–101°F). This value will rise after exertion and if the horse is fevered. A fall in temperature may indicate shock. The resting pulse rate normally lies between 36 and 42 beats per minute. This value will rise with exertion, fear, excitement and fever. The resting respiration rate is usually about 8 to 16 breaths per minute. As before, this value will rise with exercise, excitement, etc. Respiratory disease may also cause the horse to have an abnormally high breathing rate.

If the trainer has any doubt about a horse's health, the veterinary surgeon can be asked to take blood tests. The results of a blood test, while not infallible, can be useful in helping the vet to diagnose illness and assess the horse's fitness. The meaning of the test results are discussed later.

Assuming that then horse is healthy, fitness is obtained by giving it exercise. This exercise can be either general or specific. Walking and trotting are excellent general conditioning exercises, and the horse should be walked before and after work as this starts the warm-up process and increases the circulation of blood throughout the body. The hoof is a specialised structure designed to pump blood by the pressure of the horse's weight on the foot, thus pre-exercise walking improves the blood circulation in the legs and prepares them for the stress of faster work. After exercise, walking helps to decrease the blood circulation slowly so that the horse is able to recover from the stress of exercise more efficiently.

Horses intended for specific performance will need special exercises to develop the muscles and associated structures used most strenuously; for example, the showjumper needs a strong back and hind-quarters and the racehorse needs special attention to the legs. Also, specific exercises must be used to develop weak areas which may be due to injuries or conformational defects.

When exercise is combined with the correct feeding of a balanced diet, the horse will gradually become more and more fit, until it is capable of dealing with the task demanded of it. However, training is not that simple as each horse is an *individual* and that individual is constantly changing both physically and mentally. The appearance and temperament of a horse will be influenced by day-to-day variations in its environment. Every day is different and presents a different challenge.

The horse is not a machine and cannot be expected to behave like one. This means that in order to obtain full realisation of a horse's potential the trainer must be able to understand and appreciate the limitations and capabilities of that horse in the various situations with which he confronts it. The unique characteristics of a horse can influence how it reacts to and accepts a training programme, and these characteristics must be considered when formulating and carrying out such a programme.

Some of the factors that will influence a programme include:

Age. Younger horses, particularly those that have not yet reached maturity, are in many cases not physically able to withstand the stress of rigorous training, yet younger horses require more daily exercise than older horses to stay fit.

Previous training. Once a horse has been brought to peak fitness it is much easier to get it fit again.

Length of rest. The shorter a break the horse has had, the easier it will be to get fit.

Temperament. Some horses are lazy while others are keen. The keen horse may be easy to get fit and difficult to keep calm, the the lazy horse may need more work to get fit but keep its condition and be more placid.

Soundness. If a horse has a history of unsoundness – for example, tendon problems – the training programme may have to take longer and avoid too much fast work. Conformational weaknesses must also be taken into consideration because these may predispose the horse to unsoundness.

Type, including weight, size and height of horse, must also be taken into consideration.

The competitive goal or goals will also have a great influence on the type and length of training programme chosen. Horses should not be got overfit and so become difficult to manage for the young or inexperienced rider.

Learning is an important part of training, particularly when training the young horse. The horse is a herd animal, and as one of the herd it is accustomed to being guided by the behaviour of the herd leader. The average horse does not have much individual initiative compared to many other animals and, because of this, is often uncertain how to behave when it is alone. The rider or trainer exploits this characteristic by dominating the horse, and must be *confident, consistent* and *quietly forceful*, thus putting the horse at ease. The horse is most at ease when it knows what behaviour is expected of it. If the rider or trainer loses confidence, this is immediately communicated to the horse and it, too, becomes uncertain.

To obtain the best performance from the horse demands optimum conditioning of all the body systems involved during the specific work the horse is asked to perform. A fit horse is less likely to be completely fatigued at the end of its workout – for example, a race. This is important, because fatigue contributes to breakdown.

Both the intensity and the duration of the training programme will affect the degree of fitness obtained. However, it must always be remembered that there are limits beyond which different systems cannot be pushed without risking damage. For example, there is no point in developing muscle to the point where the horse starts to develop bone and joint problems.

In simple terms, both trainer and rider have to have an indefinable quality known as 'feel': knowing horses and anticipating their reactions. Feel is largely acquired by years of experience, aided by an innate sensitivity for which there is no substitute. The object of the following chapters is to help trainers and riders attain a better knowledge of how their horse's body works and how it adapts to training. This knowledge can then be used to design training programmes that do not overstress the horse and thus have better long-term results.

Today's trainers and riders must not be blinded by science and technology, but should benefit from a mixture of horse-sense from the past and modern knowledge. Commonsense must not be lost sight of. The basic principles of horse and stable management are very important and can be summarised as follows:

(1) An airy, safe loose box with no draughts.
(2) Well-fenced, clean grazing.
(3) Fresh, clean forage.
(4) A constant supply of clean water.
(5) Thorough grooming.
(6) Shoeing.
(7) Exercise programme.

These must all be incorporated into a daily routine to get the best from the horse. The significance of each of these factors is explained later, but each is of vital importance to an optimum training regime.

2 Traditional Training Methods

The horse has been domesticated for thousands of years. It is probable that our ancestors initially herded horses as a meat animal but then realised that here was an animal that could not only cover great distances but was also capable of high speed. The horse became indispensable as a working animal and very stringent demands were made upon it in terms of speed and endurance; consequently, a high degree of fitness was needed.

Foxhunting developed in the eighteenth and nineteenth centuries to become a premier equestrian sport, and from the training of horses used for foxhunting developed the traditional methods of getting horses fit from grass which are outlined below. These are tried and trusted methods used by generations of grooms to get horses fit for the November Opening Meet, with October cubhunting supplying some of the fast work needed for true fitness. When getting horses which are not going to hunt fit, the fast work must be included in the training programme itself, and this, too, will be outlined.

Traditionally, hunters are brought up from grass on 1 August, allowing three months to get them fit enough for a hard day's hunting – the equivalent of which would be a one-day event or a 20 mile distance ride. This three-month period can be split into three four-week periods: preliminary work, development work, and fast work. It is important that adequate time is spent on each of these periods, so that a depth of fitness is gradually built up. Modern competition horses are rarely given such long rest periods at grass as is traditional for hunters and therefore may be fittened in less time.

Getting a Horse Up from Grass

Getting a horse fit involves exercise, feeding and overall stable management. Before embarking on an exercise programme, some routine procedures should be carried out which are essential to the health and well-being of the animal:

RECORD OF EQUINE VACCINATION

This document should be held by the owner/keeper of the animal described overleaf and must be presented whenever proof of identity is required. Please keep the document in a safe place as failure to produce it may result in the animal being debarred from a showground, stables or event site.

The document must be passed to the new owner when the animal is sold.

The Veterinary Surgeon carrying out a vaccination should check the identity of the animal against the diagram and description overleaf before signing the form.

Name of owner	Address of owner

N.B. This record is not for use under The Rules of Racing or The Conditions of the General Stud Book.

VACCINATION

Name of Vaccine and Batch No.	Date	Vet. Surgeon's name and address (block capitals or stamp) and signature.
PREVAC-T 018N05	12/7/81	
PREVAC-T 175	7/8/81	
PREVAC-T A180	2/8/82	
DUVAXYN-IE-T UT 01901	1/8/83	
DUVAXYN-IE U026001	17/7/84	
PREVAC-T A002	16/7/85	

Fig. 2.1 Vaccination record

Shoeing. The farrier should be called to check the horse's feet and fit a new set of shoes. Stud holes will probably not be necessary at this stage. New shoes should be fitted every four to six weeks while the horse is in work.

Teeth. The teeth should be checked for sharpness and unevenness. Any sharp edges will need rasping by the vet or horse dentist so that the horse does not experience discomfort when eating or when the bit is in its mouth. Young horses (up to five years old) should also be checked for the loss of milk teeth and the eruption of permanent teeth which may cause pain and hence evasions. Wolf teeth may also cause problems and should be removed if troublesome.

Vaccination. Horses should be injected annually, before a rest period, against equine influenza and biannually with an anti-tetanus booster. (Fig. 2.1 shows a detailed vaccination record.) The tetanus status of all horses in the yard should be known so that should an injury occur the vet will know whether or not a tetanus injection is necessary. Influenza vaccinations are compulsory for many competition horses; they are also important for the hunter which encounters large groups of strange horses at least once a week. Competitors should note that to gain entry to a racetrack stables the vaccination requirements are more stringent than for a showground and the certificate is checked for intervals between vaccinations right back to the primary injection.

Worming. Horses should be regularly wormed (fig. 2.2) whether they

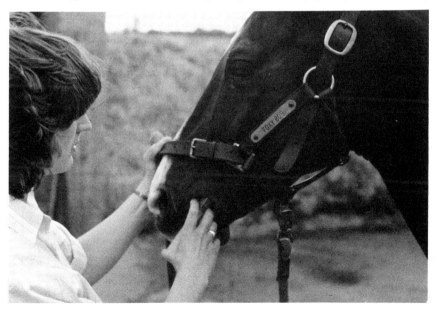

Fig. 2.2 Worming

are out at grass or stabled. It is very unfair to expect the horse to perform strenuous work while suffering from a worm burden.

Checking. The horse must be checked over very thoroughly every working day for bumps, swellings and scratches. Any injury, no matter how trivial, should be noted and treated. Particular attention should be paid to the lower leg and hoof, looking for signs of heat, swelling and tenderness.

Bathing. If the weather is reasonably mild the horse can be washed (fig. 2.3) from time to time during the training programme. This will help rid the coat of parasites, grease and scurf which may be making the horse feel itchy. It will also improve the horse's appearance. A hot, sweating and blowing horse should not be washed as horses are susceptible to chills; nor should a horse be washed unless it can be thoroughly dried. Soap or shampoo and plenty of warm water should be used all over the horse's body except for the head which is washed last with clear water. Rinsing should be thorough, all traces of shampoo being removed as any left may cause a skin irritation. Excess water is then removed with a sweat scraper and the horse rubbed as dry as possible with a towel. Finally, the horse should be walked dry with a sweat rug on.

A hunter brought up in August can be kept on a 'half-and-half system'

Fig. 2.3 Bathing

for the first few weeks. This involves either keeping the horse in the stable during the day (to avoid the heat and flies) and turning out in the paddock at night, or just bringing in at night. Obviously, this is not practical for a horse brought into work during the winter. There are several advantages to the half-and-half system:

(1) It is always preferable to make any changes in the horse's routine gradually, allowing the digestive system to adapt to the hay and corn ration.
(2) A horse will be less exuberant if he spends some time in the field instead of shut in a stable.
(3) Young horses particularly may get very bored or upset at suddenly being confined in the stable for 23 hours a day.
(4) Thin-skinned Thoroughbred horses may actually lose condition out at grass if harassed by flies.

One disadvantage is that once horses have had their hind shoes back on it is unwise to turn them out together: even an irritable kick against flies may cause a nasty injury on another horse.

Preliminary Work

The preliminary work is designed to exercise the horse slowly for increasing lengths of time, and is accomplished by road work at the walk. This work tones up the muscles, tendons and ligaments and is known as 'hardening the legs'. While not physiologically correct, this term sums up the role of preliminary work.

Walking work is alleged to begin the process of turning the fat, acquired out at grass, into muscle. Strictly speaking, this process does not happen. The muscle system gradually develops; simultaneously, the fat reserves of the horse are 'burned up' to provide energy for work. This process is more noticeable after faster work begins.

The first few days' work can be done using a mechanical horse walker, long reining or by ride-and-lead. Commonly, with horses of reasonable temperament, riding is commenced from the outset. Any lunging should be undertaken with care as the horse's system is not ready for possible stress which can be created in small circles.

Initially, it is best to start with 30 minutes a day on flat, easy ground, building up by 10 to 15 minutes a day until at the end of the first week the horse is walking for one hour. By the end of the second week, the horse should be up to two hours' walking work a day, including some hill work if possible. During the third week, some gentle trotting uphill can be included. Uphill work exercises the muscles and wind to a greater extent and has a less jarring effect on the front legs. The routine is much the same during the fourth week. Trotting periods can be longer, but never so long that the horse blows and sweats.

Do not be tempted to cut the preliminary work short, particularly the walking work, no matter how boring it may become. It is essential to develop and harden the horse's muscle and reaccustom the horse to carrying a rider without overstressing the heart, lungs or legs. *It is impossible to overemphasise the importance of this walking period.* If early exercise is not done gradually and thoroughly, the risk of a horse becoming lame is much greater.

During these walking sessions the horse must be active and working at the walk. On the other hand, a fresh horse straight up from grass may misbehave, jogging and pulling. The horse must be made firmly and quietly to walk. These high spirits must not be 'worked off' at a faster pace.

Throughout all training, every aspect of horse and stable management is important; and it is particularly so at this time. For example, during the first fortnight the horse must be carefully checked every day for rubs, galls and injuries. The horse's skin is soft after its summer rest and is easily irritated by ill-fitting or badly cared for tack. Areas most at risk are the corners of the mouth and the girth and saddle region. A thick numnah and a soft girth or a soft girth-sleeve should be used. The saddle and girth areas can be hardened by bathing with salt water, methylated spirits, surgical spirits or witch-hazel. An added precaution is loosening the girth (not too much!) for the last 15 minutes of a ride, thus allowing the skin under the girth to cool gradually. When tacking up, the horse's front legs should be pulled forward to stretch the skin into a comfortable position under the girth. Girths and numnahs should always be kept clean, being brushed or cleaned after use. Sponge-filled numnahs appear to cause sore backs more easily than others, just as sponge-lined boots irritate legs.

As the majority of this work is roadwork, knee boots may be worn to protect the unfit horse which may be prone to stumbling. An unfit, weak-muscled horse may knock itself, particularly when starting trotting work. A careful eye should be kept on the legs, and brushing boots or Yorkshire boots fitted if necessary.

Generally speaking, trotting should not be introduced until the horse is returning to the stable after a two-hour walk as fresh as it left it. Trotting is very jarring to the horse's legs and initial trotting periods should be slow but active, short and preferably uphill. Four 200 metre trots would be adequate the first day. The horse should have walked for at least 15 minutes so that it is warmed up before the first trot.

After a few days' trotting, lunging can be introduced into the programme if desired. Initially, this should be not more than six or seven minutes on each rein before or after the ridden work. Ridden schooling work can be included during the fourth week, consisting of large circles in walk and trot on both reins. Lunging and schooling will help prevent the horse getting bored or stale. Boots should be worn all round during this sort of work.

Development Work

By the end of the fourth week of roadwork, with some uphill trotting towards the end of the period, the horse should be ready to go on to the next stage of training and some cantering and suppling exercises can be incorporated. Horses will vary in the time taken to reach this stage. The best guide is how well the horse is coping with the uphill trotting: if its breathing at the top of the hill is fairly even and not strained, it is probably time to start canter work.

Development training will vary, depending on which sport the horse is being got fit for. The polo pony will be given some basic schooling and stick-and-ball work; the showjumper will be asked to jump small fences and combinations of fences; the racehorse will be cantering; the long distance horse will step up distance and speed; and the event horse will be given suppling exercises on the flat and over very small fences. All these aspects will be considered in separate sections later in the book. The traditional training methods were developed for the hunter, and it is the hunter that will be considered now.

The ground chosen for the first canter should be flat and good to firm, not soft and dotted with pot-holes. The overexuberant horse may be tempted to buck and go faster than desired, risking leg injury, so the rider should be prepared for this. Ask quietly for canter, and keep the horse's head up. If the horse is known to be strong, it should have on suitable tack. Canter work should be slow and short initially and gradually increased so that by the end of the sixth week three or four periods of *steady* cantering can be included in the exercise programme. At the end of each workout the horse will be blowing, but this should not be excessive and the horse should recover from it quickly during the trotting which should follow fast work.

The exercise routine continues in much the same vein through the seventh and eighth weeks. This work is building up the horse's stamina (heart and lung endurance), with some uphill canter work being very beneficial. During this time (mid to late September) the horse is ready to go cubhunting one or two *short* mornings a week. Make sure that the horse is not kept out too long, initially. When starting at 7 or 8 a.m., it is very easy to stay out until lunch time and find that the horse has actually had a longer day than if it had been hunting. Although the horse may not have galloped and jumped, the number of hours it has been out of its box with a rider on its back will be very tiring for the semi-fit horse.

Fast Work

The hunter is rarely given any fast work during its exercise programme, October cubhunting usually providing the occasional pipe-opener which prepares the horse for the beginning of the season. It is generally felt that the hunter hammers its legs quite hard enough from November to March/April out hunting, without galloping it in between hunting days! It is a good idea to school the horse over some cross-country or natural fences once or twice during October. The hunter may not have jumped a fence for five months, and it may avoid embarrassing moments to refresh the horse's memory.

Following this sort of programme, the hunter should arrive at the Opening Meet ready to run for its life with a depth of fitness suitable to see it through the ensueing five-month hunting season.

During the Season

Throughout the hunting season, days of sport will be interspersed with exercise days. Exercise on these days will follow much the same pattern as the exercise given during the preliminary part of the training programme. The exercise given will, of course, vary depending on how hard the horse is hunted. A hunt horse may do no more than one long, hard day a week or two half days a week, the latter being preferable. This horse will be kept fit by the hunting it does and other exercise should be no more than 60 to 90 minutes' walking and a little trotting to keep it ticking over. Sunday will probably be a rest day, with the horse walked out to graze for 10 minutes in-hand. The day following hunting, exercise may be 30 minutes' walking to relieve any stiffness.

A subscriber's horse may only hunt one day a week, that day being considerably less arduous than the hunt horse's day. This horse would have a short walk the day after hunting (usually a Sunday) and possibly a day off mid-week when it would be walked out in-hand for 10 minutes or so. The remaining four days of the week it should have 60 to 90 minutes' exercise, including walking, trotting and cantering as during the development phase of the exercise programme. This horse may also benefit from a short 'pipe-opener' the day before a hunting day, to clear its wind and prepare it for the next days' galloping.

Horses may hunt twice a week, three days a fortnight or once a fortnight, and the exercise given to the horse between hunting days will vary accordingly. Horses hunting two days a week should be treated along the same lines as the hunt horses. They are working very hard and will need the minimum of exercise to keep them fit. The horse hunting once every two weeks should be given some fast work once or twice during the non-hunting week or it will never reach peak fitness. This fast work could perhaps involve half to three-quarters of a mile (750 to 1000 metres) at half speed in week 10 of the training programme, building up to one mile (1500 metres) at three-quarters speed by week 15 of the programme (mid-November).

Barring accidents, a properly fed and exercised horse can stay at peak fitness throughout the whole of the hunting season, whether it is hunting two days a week or one day a fortnight. All horses will vary in the exercise needed to keep them in tip-top condition according to their temperament, type and the work they are doing. These factors must all be considered when planning the exercise regime for a hunter. The complementary aspects of nutrition and good management are discussed in subsequent parts of this book.

The staghunting season lasts for nine months of the year, which means that there is not actually enough time to let a horse down, give it a complete rest and get it properly fit again. If one is to hunt the whole season, there are two alternatives. The first is to have two horses and treat them like foxhunters, working each for half the season. This, of course, may be too expensive, and the other alternative is not to rough the horse off during the three months non-hunting but keep it ticking over so that it never looses its fitness and yet relaxes and gains condition. Many staghunting people hunt their horses off grass, supplemented with corn during spring and autumn hunting, which means that horses are more relaxed and keep their condition better than fully stabled hunters. The exercise programme will have to be adapted to suit each individual horse's and rider's requirements.

Roughing Off

At the end of the hunting season the hunter is given a long summer's rest, traditionally being turned out to grass on 1 May. The hunting season seems to be finishing earlier and earlier due to modern farming pressures, and most Hunts now having their closing meets around March or April time. A clipped-out, corn-fed hunter should not be suddenly turned out at this time of the year when the weather is, to say the least, unpredictable and the grass growth uncertain. A gradual 'letting down' process should be followed, heading towards the resting period. Exercise should be reduced and, simultaneously, the corn ration cut down and the amount of hay fed increased. The number or weight of rugs should be decreased as the weather gets warmer, and the horse should be turned out into the paddock for a few hours a day as soon as the ground can stand it. At this point exercise can stop and the hind shoes can be removed. It is wise to keep front shoes on to prevent foot problems. The horse's summer coat should start to grow and it should start to look less lean and hard. By the end of April the horse should be out all day even if it is still coming in at night. It is worth taking time and trouble over the roughing-off period or the horse may get cold and lose condition which may take a long time to recover.

During June and July the weather may be very hot and the horse may be pestered by flies, causing loss of condition, particularly in thin-skinned Thoroughbred types. It may be worthwhile stabling the horse during the day and turning out at night if adequate shelter is not available. During the summer rest period, the horse must be wormed every four to six weeks, and checked daily for minor injuries. The farrier should trim the feet and front shoes, if kept on, must be checked and changed when necessary. The field must be checked regularly to ensure that the fencing and water supply are safe and that there is no sign of litter or poisonous plants or, indeed, any other hazard to the horse.

Having got a horse to peak fitness during the hunting season, it seems a shame in many respects to let it get soft and flabby. Human athletes do not get gross for three months of the year and then struggle to get fit again. They may let themselves have a 'rest' by easing off slightly, but they always keep themselves ticking over. It puts a tremendous strain on the body's systems, heart, lungs, muscles and tendons to get unfit and then have to get fit again.

However, horses that have hunted hard all season definitely need to rest and relax. A good compromise is to rough the horse off and turn it out but keep its shoes on and exercise it once or twice a week. This exercise could be 60 to 90 minutes' walk and trot, enough to keep the horse's back and girth area used to the saddle and the body systems used to working. This system may not suit everybody, but is much healthier for the horse.

Part II

How the Horse Works

INTRODUCTION

The horse's body can be thought of as being made up of several integrated systems, each system consisting of organs and tissues specialised for the specific tasks of that system. The systems are:

(1) The respiratory system.
(2) The circulatory systems of both blood and lymph.
(3) The locomotive systems of skeleton, muscles and sinews.
(4) The systems of information and control; nervous, sensory and endocrine systems.
(5) The dental system.
(6) The skin.
(7) The reproductive and mammary systems.
(8) The urinary system.
(9) The digestive system.

The first four systems are directly involved every time the horse is asked to work, and the respiratory, circulatory and locomotive systems will be considered in detail.

The dental and digestive systems have incidental effects on the fitness of a horse. If its teeth are sharp it will not eat properly, and if it fails to make good use of its food it will not thrive. A vital part of training a horse is feeding it correctly and the digestive system and nutrition are dealt with in chapter 3.

The skin is involved in heat loss and sweating which is vital to temperature control during exertion.

It can be seen, therefore, that to get the best from the horse the trainer must understand the main systems and their functions.

3 Nutrition of the Performance Horse

The Importance of Feeding

Feeding is one of the most critical factors in determining the performance of the horse. The horseman relies on the experience of generations of tried and tested feeding regimes, always allowing that horses are notoriously individual in their needs. The aim of this chapter is to extend this experience by combining the highest traditional standards of feeding with sound scientific knowledge, and to explain away some of the myths of feeding from a scientific point of view without losing any practical and proven methods.

It might be argued that the methods used by our grandfathers are quite adequate for today's horse without complicating the matter with scientific mumbo-jumbo. However, there are several reasons why the horseman should be increasingly aware of the nutritional needs of his horse:

(1) Top-class competition horses are subjected to high levels of stress when they are travelled great distances by land, sea and air to competitions.
(2) Combined with this, there is prolonged close contact between strange horses when stabled at competitions which can lead to subclinical disease.
(3) Feed production technology has changed, involving processing methods which may affect nutrient levels and availability.
(4) Unusual, cheaper protein sources are included in compound feeds and may stimulate an allergic reaction such as urticaria or hives.
(5) Intensive farming methods may deplete the soil of its nutrients, leading to deficiencies in the crops grown in that soil.

There is increasing anxiety amongst horsemen as the results of these factors show up as training problems. A greater understanding of the nutritional needs of the horse will lead to the formulation of better feed rations and hence healthier horses.

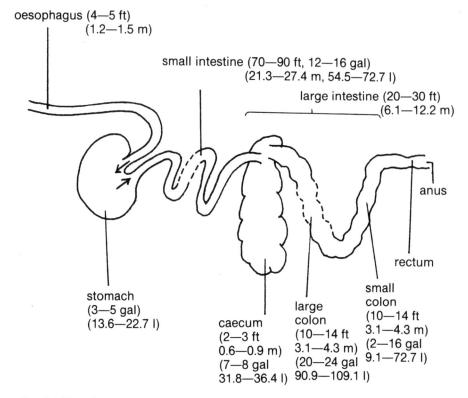

oesophagus (4—5 ft)
        (1.2—1.5 m)

small intestine (70—90 ft, 12—16 gal)
        (21.3—27.4 m, 54.5—72.7 l)

large intestine (20—30 ft)
        (6.1—12.2 m)

anus

rectum

stomach
(3—5 gal)
(13.6—22.7 l)

caecum
(2—3 ft
0.6—0.9 m)
(7—8 gal
31.8—36.4 l)

large
colon
(10—14 ft
3.1—4.3 m)
(20—24 gal
90.9—109.1 l)

small
colon
(10—14 ft
3.1—4.3 m)
(2—16 gal
9.1—72.7 l)

Fig. 3.1 The alimentary canal of the horse, showing dimensions and capacities of different regions

The Equine Digestive System

The horse is essentially a grazing animal and has developed throughout evolution as a 'trickle-feeder'. The gut is designed to cope with the regular intake of small quantities of fibrous food and it needs a constant supply without ever overloading the system.

The Facts Behind the Rules

A stabled horse is in an artificial environment, totally dependent on us for his food and water. In order to keep the horse's system working efficiently, the feeding regime must mimic nature as closely as possible, hence the first rule of good feeding: *feed little and often*. The key reason for this is that, for its size, the horse has a relatively small stomach – about the size of a rugby ball (fig 3.1). The stomach can stretch to accommodate about 13 to 23 litres (3 to 5 gallons) of food, but, due to the 'J' shape of the stomach it is never more than two-thirds full, i.e. the stomach will hold

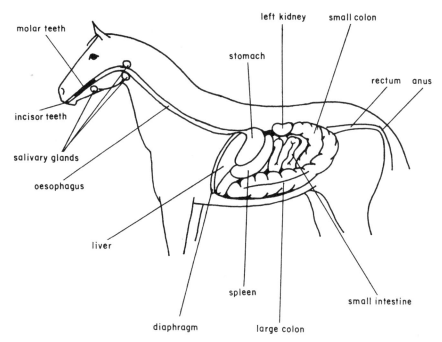

Fig. 3.2 The digestive system viewed from the side

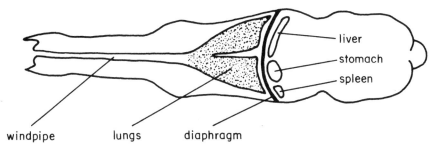

Fig. 3.3 The relationship between the stomach, diaphragm and lungs

about 9 to 13 litres (2 to 3 gallons) of food, about two-thirds of a standard water bucket.

Allowing for about one-third of the stomach's contents to be water and saliva (the horse produces large quantities of saliva), the amount of food most usefully consumed in one meal is limited to half a bucketful. Any food fed over and above this – for example, a bucketful of hard feed fed to a greedy horse – will push partially digested food out of the stomach before the feed has been finished. Not only is this wasteful; it is potentially dangerous, too, since this partially treated food may ferment in the small intestine and cause colic.

A swollen, full stomach will also put pressure on the diaphragm (the muscular layer separating the lungs from the guts), preventing the horse from filling its lungs properly – hence the rule *do not work fast immediately after feeding* (figs. 3.2 and 3.3). Nobody wants to run a cross-country after Sunday lunch – the horse is just the same!

Another factor to be considered is that when the horse works some of the blood supply is diverted from the gut to the muscle. Digestion will slow down but food will still pass along the gut, resulting in the horse getting fewer nutrients out of the food before it is expelled as faeces. This is also the reason why horses and humans may suffer diarrhoea (scouring) after severe exertion.

When the stomach is full and the horse takes a long drink of water, food may be washed out of the stomach before it has been treated by the gastric juices in the process of digestion. As before, this leads to wastage and possibly colic. So the rule *water before feeding* is soundly based. However, the extent of this 'washing out' is debatable and modern thinking is that the effect is not as serious as once believed. Due to the 'J' shape of the stomach, it may be that water merely washes over the top of the food causing little harm. To avoid any problems the horse should *always have access to fresh clean water* except when just about to compete at high speed.

The size and nature of the caecum and large intestine (fig. 3.1) indicate the horse's need for fibre in the diet. There should never be less than 25% roughage in the daily ration, so *feed plenty of roughage*.

Problems can arise in competition horses which compete with a gut full of roughage, as the roughage is bulky and requires large quantities of water in its digestion. Thus a high roughage diet gives the horse a big belly and will amount to it carrying an unnecessary burden. The extra weight and the pressure on the diaphragm will also impair heart and lung function, yet the horse must eat at least a quarter of its daily intake as roughage in order to keep the gut working properly. Sugar beet pulp is a useful source of highly digestible fibre, and will be discussed later.

The digestion of roughage takes place in the caecum and large intestine, making available to the horse nutrients not available to carnivores, such as dogs, and omnivores, such as humans – we would not look very well on a diet of grass. The process that takes place in the caecum and colon is called 'fermentation' and is carried out by a vast population of micro-organisms (bacteria and protozoa). These specialise in fermenting particular parts of the diet and so the number of each type of organism present in the gut will vary depending on what the horse is being fed – sudden changes in the diet will result in there not being sufficient numbers of the appropriate organism to deal with the food, and food will pass through the gut only partially digested. This is at best wasteful and at worst may cause diarrhoea; therefore *make any changes in the diet gradually*. This is partly the reason for the laxative properties of a once-weekly bran mash.

The horse is susceptible to problems such as *laminitis* and *azoturia*, caused in part by overfeeding, so it is vital that *the food level anticipates the work load*. Days when the horse is off work must be planned and the ration altered accordingly the day before. Horses laid off due to illness, lameness or bad weather should have their corn feed cut dramatically. If rest days cannot be anticipated, the feed on the day should be reduced in concentrate content and every effort made to either turn the horse out or lead it out in hand.

The horse must be fed a diet which is *balanced and suitable for that individual and the work it is expected to do*. This is, of course, easier said than done and will be discussed further. An overfed horse can be dangerous, to itself and to its rider.

The remaining 'rules' are self-evident, and one of them is *always use good-quality feedstuffs*. There is an increasing awareness of the horse's susceptibility to dust allergy which can only be avoided by feeding clean, good-quality forage. The horse's appetite will be depressed by stale food or dirty mangers, and therefore *scrupulous attention must be paid to hygiene and storage facilities*. Certain foods have a limited shelf life, beyond which they may lose palatability and nutritive value, and this must be considered when buying feeds or supplements in bulk.

The horse is a herd animal and a creature of habit, and benefits from having *regular feeding hours every day*. A horse may fret if its feed is not available at the expected time and thus lose condition, which will affect its performance.

The bacteria in the horse's gut are able to synthesise some vitamins but *feeding succulents* is an excellent way of tempting appetite and providing a natural source of vitamins and minerals.

These 'rules' have been adhered to for many years, since long before the anatomy and physiology of the horse's gut was understood. Our increased understanding of what takes place in the horse's body has revealed that these rules are firmly based on scientific fact, which is why they have stood the test of time so admirably.

The Nutrients Needed for a Healthy Horse

In order to remain in peak condition, the working horse requires a regular supply of about forty nutrients, and to obtain these nutrients the horse has to eat suitable food. Nutrients fall into seven broad categories: carbohydrate, fat, protein, fibre, water, minerals and vitamins.

Carbohydrates

Carbohydrates include substances such as sugar, starch and fibre, which make up the majority of the horse's diet and are the horse's major source of heat and energy.

Sugars are the simplest forms of carbohydrate, the basic building blocks being monosaccharides or simple sugars, e.g. glucose. It is in the form of glucose that carbohydrates are absorbed through the walls of the small intestine. Glucose is then stored in the horse's body as *glycogen* and fat which can be converted back to glucose to be used by the muscles during exercise. In excess, carbohydrate can cause problems such as laminitis, lymphangitis and azoturia.

Starch is the major energy store in plants. For example, roots and tubers contain 30% starch and cereal grains about 10.5% starch, making them both very useful sources of carbohydrate for the horse.

Cellulose or *fibre* (roughage) is a complex carbohydrate which is very important in the structure of plant cell walls. Cellulose is broken down by the horse in the caecum and large intestine, and is a very important source of energy in the grass-fed horse. Lignin is a type of fibre which is resistant to breakdown. This is significant because the lignin content increases as the plant ages, and thus mature plants are less digestible (i.e. they are less nutritious). However, fibrous roughage is an essential part of the horse's diet – plenty of bulk food is needed to aid breakdown and digestion of other feeds and to maintain the health of the gut. At least 25% of the total diet must be provided as roughage.

Fats and Oils

Fat is a very concentrated form of energy, and provides more than twice the heat and energy per gram of carbohydrate. Fat can be stored at a greater concentration in the body cells to form more permanent stores of energy; it also provides insulation as subcutaneous fat. Until recently, fats have been something of an unknown quantity in horse diets, most traditional rations containing about 2 to 3% oil. It is possible that greater use of fat as an energy source may help reduce problems associated with feeding too much carbohydrate – for example, azoturia and laminitis. Some feed manufacturers are including up to 6% vegetable oil in their high-performance horse feeds, and some nutritionists are recommending the inclusion of fat in the diet of endurance horses.

Protein

Protein is concerned with the replacement of muscle tissue lost through natural wastage and the building up of new body tissue. When there are excessive levels of protein in the diet, it can also be used as a source of energy.

Protein is absorbed across the small intestine wall as amino acids, which are the building blocks of body tissue. There are about 23 different amino acids, of which 10 are 'essential' and must be included in the diet. The remainder can be synthesised by the micro-organisms already in the caecum. This means that the proteins in the diet must not only contain the right types of amino acid; the amino acids must also be present in the correct balance, i.e. the horse needs high-quality protein. The quality is reflected by the 'biological value' of the protein, which shows the number of essential amino acids present. Animal protein (e.g. bone meal) has a high biological value, but plant protein (e.g. oats and barley) tends to have a low biological value. For example, the amino acids lysine and methionine may be lacking in horses fed a traditional diet of hay and oats, which is why our grandfathers used to include other ingredients such as beans.

The protein levels in the horse's diet must compliment the energy content in the ration since optimum feed utilisation depends on a correct protein to energy ratio. This means that although hard work only requires a small increase in protein to make up for a slightly quicker muscle turnover, protein levels must increase as the energy content of the ration increases to maintain the protein to energy ratio. However, feeding very high protein levels to competition horses is wasteful and expensive, because excess protein is broken down and used as an alternative energy source. Overfeeding of protein has even been shown to be detrimental to performance.

Water

Water is a vital component of the diet and plays an essential role in nutrition. Nutrients must pass into the horse's system as a solution. Water is needed for this, and is also necessary to produce the large amounts of saliva which help swallowing. A horse can lose nearly all its body fat without major ill effect, but a loss of only 8% of the body water causes illness and a loss of 15% can cause dehydration and heat stroke in competing horses.

Water intake will be affected by several factors including diet, environmental temperature and the amount and type of exercise the horse is undergoing. A lack of water will adversely affect feed consumption, a fact which may be important when caring for the sick horse which is reluctant to drink.

Clean, wholesome water should be freely available for horses at all times. If this is not possible – for example, during work – and the horse has become excessively thirsty, it should be offered small amounts of water at intervals until its thirst is quenched.

Minerals and Vitamins

The pages of any horsey magazine are covered with advertisements for supplements, each one claiming to be the best for your horse. How is the horseman to decide which to buy? The object here is not to recommend a particular supplement but to show the horseman what to look for in a supplement by describing the job that minerals and vitamins do and indicating where deficiencies may arise. A supplement is not a wonder drug: its task is to make up the difference between shortcomings in the diet and what the horse should be receiving. A supplement, when added to the diet, corrects any imbalance of nutrients whereas an additive is merely an additional ingredient acting, for example, as a flavouring.

Minerals

There are about fourteen essential minerals which have been shown to be vital to the biological processes which take place in the body. These include the macrominerals required in larger amounts (calcium, phosphorus, sodium, chlorine, potassium, magnesium and sulphur) and the trace minerals required in small amounts (iron, copper, iodine, cobalt, manganese, zinc and selenium). Other minerals such as molybdenum and fluorine may also play an important role, but this has not yet been fully established.

All natural foods contain mineral elements, the proportions varying greatly between different types of food, e.g. cereal grains are low in calcium and magnesium and high in phosphorus, while forages contain more calcium and magnesium but less phosphorus. Such foods can only contain the minerals present in the soil in which they are grown, and this can lead to deficiencies in certain areas. Deficiencies also arise as a result of the intensive nature of modern agriculture, which gradually depletes the soil of certain minerals. Horsemen should be aware of any local soil characteristics which may affect the composition and quality of feedstuffs.

The mineral elements which may be inadequate in traditional diets include calcium, phosphorus and magnesium. Sodium, potassium and chloride (sodium chloride is common salt) may be deficient following prolonged exercise as they are lost in sweat. This loss may be made good promptly by the use of electrolyte solutions during and following competitions. The most frequently detected trace element deficiencies are selenium, zinc, manganese and iodine.

Diagnosis of deficiency is difficult because there are no specific symptoms. Borderline deficiencies may exist where the horse shows no external signs of illness but is just not performing as well as it should. A competition horse should receive a daily supply of minerals and vitamins so that the body has the chance to function properly. This becomes even more important when the horse has only limited access to good quality pasture and is under the stress of top competition.

Calcium and Phosphorus

These minerals are essential for the formation and maintenance of healthy bone, and their effectiveness is dependent on an adequate supply of vitamin D.

Mineral and vitamin requirements are greatly influenced by interrelationships; not only must the minerals and vitamins be supplied in the daily ration, but they must be supplied in the correct amounts and proportions. This is of particular importance with calcium and phosphorus – too little of either one limits the usefulness of the other. Adequate quantities in a ratio of 1:1 to 2:1 (calcium to phosphorus) are essential for proper utilisation of the minerals in the body. The majority of cereals fed to horses have a very poor ratio, and horses on a high corn ration and growing young stock should be supplemented with limestone flour or bonemeal for extra calcium. Bran inhibits calcium uptake, so when bran is used regularly a calcium supplement is needed.

Magnesium

This element is also vital for proper bone and tooth development, and is usually adequately supplied in the diet.

Sodium and Potassium

Sodium chloride (common salt) and the potassium salts help control fluid balance in the body and play an important part in blood formation and food digestion. Deficiency shows as tiredness, particularly in high performance horses which sweat profusely. It is helpful if 30 g (1 oz) of common salt is fed daily in the feed.

Iron and Copper

Iron and copper are essential for the formation of blood pigment (haemoglobin) which is responsible for the carriage of oxygen in the blood.

Manganese and Zinc

These help activate the enzymes (biological catalysts) which break up the food during digestion. They are also important in maintaining the health of skin and coat. A lack of zinc is rare, but it will depress appetite.

Iodine

A deficiency of iodine may be the cause of a horse looking off colour and listless, because iodine is an essential part of the hormone thyroxine which governs the rate of body metabolism.

Cobalt

Cobalt is a component of vitamin B_{12} and is used to help combat anaemia.

Selenium

In conjunction with vitamin E, selenium is thought to help prevent cell damage – particularly muscle tissue – and may be used to treat azoturia. Some nutritionists recommend selenium and vitamin E as supplements for the diet of high-performance horses.

Vitamins

Vitamins are vital components of the diet, and are essential for normal metabolic functions and optimum utilisation of food. They are only required in tiny quantities and are known as 'micronutrients'. A traditional ration of oats and hay is likely to be deficient in some of these vital nutrients. However the bacteria in the caecum are able to make certain vitamins, the level of synthesis depending on the vitamin involved and the type of ration being fed. The amount of vitamins absorbed from the large intestine to be used by the body is low and synthesis can be disrupted by external factors, e.g. antibiotic therapy.

Vitamins fall into two main categories: the fat-soluble vitamins A, D, E and K which can be stored for a limited time in the liver, and the water-soluble vitamins C and the B complex which cannot be stored.

Vitamin A

Necessary for healthy bone and tissue development, vitamin A is found in grass, green foods and good-quality hay. It can be stored in the liver, but these stores can be exhausted during the winter and a supplement, e.g. cod liver oil, may be needed.

Vitamin D

This is also known as the 'sunshine vitamin' because it can be made in the skin when it is exposed to sunlight. Vitamin D is needed for the gut to be able to absorb calcium and phosphorus, and a deficiency can cause bone problems. Again, it is stored in the liver, but these stores are generally used up by mid-winter and cod liver oil provides a good supplementary source.

Vitamin E

Vitamin E is also known as the 'fertility vitamin', and is important in muscle function. When used together with selenium, it has been reported to improve the performance of horses in hard work and to combat the effects of azoturia or tying-up.

Vitamin K

This vitamin is essential in the process of blood clotting and is sometimes included in the diets of horses prone to nosebleeds. A deficiency is rare because the micro-organisms of the gut can make vitamin K.

Vitamin B Complex

The B vitamins (thiamin, riboflavin, nicotinic acid, pyridoxine, panto-thenic acid, biotin, folic acid, B_{12} and choline) are found in fresh herbage and protein-rich foods, and are also synthesised by bacteria in the gut. They are involved in carbohydrate utilisation and help the horse make the best use of the nutrients it eats. Horses with a very high carbohydrate intake may need higher levels of B vitamins, as may horses on antibiotic therapy and youngsters.

Vitamin B_{12} is essential for protein metabolism, and hence growth and reproduction. A deficiency can cause anaemia, but is unlikely in horses.

Thiamine is said to have a quieting effect. Folic acid deficiency causes nutritional anaemia because it is involved in the synthesis of red blood cells. Biotin has recently been shown to be involved in hoof growth and horn quality.

Vitamin C (Ascorbic Acid)

As the gut microbes are able to synthesise vitamin C, a deficiency is unlikely. However, it is given to horses to help them withstand stress, recover from anaemia and prevent nosebleeds.

Broadly speaking, a healthy horse receiving a good quality diet will need supplementation with salt, calcium, and the amino acids lysine and methione which tend to be lacking in cereals. Vitamins A, D and E and folic acid may also be necessary, especially during the winter months. If a horse has a particular problem, e.g. anaemia or chronically bad feet, it may be considered necessary to feed a specialised supplement and it is likely that veterinary advice would be sought.

What to Feed the Performance Horse

The task of the digestive system is to convert nutrients into end-products which can be absorbed across the gut wall. These end-products pass into the blood stream and are used by the body, firstly to stay alive, a process known as 'maintenance', and secondly for 'production', which includes any function over and above maintenance, e.g. pregnancy, lactation, galloping and jumping, or simply growth.

If the horseman is to feed his horse so that it can perform the work being asked of it, he must be able to match the nutrient requirements with what is actually contained in the foods available.

Types of Feed and their Nutrient Value

Feeds can be analysed to indicate the nutrients they contain, and a basic knowledge of the composition of various feeds is essential when preparing a feed ration for a horse. Every horseman is aware that oats and beans are more 'heating' than barley, but this superficial knowledge is not adequate and tables showing feed values should be used.

The types of feed given to horses fall into four main categories:

(1) *Cereals.* The energy-giving basis of the concentrate ration, e.g. oats, barley, and maize.

(2) *Protein feeds.* These can be of animal origin (e.g. bone meal and dried milk) or of plant origin (e.g. linseed, soya and other beans and peas).

Table 3.1 The nutritive values of some common feedstuffs

	Crude protein %	Oil %	Fibre %	Digestible energy MJ/kg dry matter	Calcium %	Phosphorus %	Dry matter %
Cereals							
Oats	11	5	12	14	0.1	0.4	
Feed barley	11	2	5	15	0.05	0.4	
Maize	10	4	2	17	0.01	0.3	
Protein feeds							
Bonemeal	27				30	10	
Dried milk	36	0.5		17	1	0.8	
Linseed	26	39	6	27	0.3	0.6	
Soya bean meal	50	1	6	17	0.25	0.6	
Bean	25	1.5	9	15	0.2	0.7	
Intermediate feeds							
Wheat bran	17	4	12	12	0.8	3	
Sugar beet pulp	10	0.5	15	14	0.6	0.07	
Compound feeds							
* Horse and pony cubes	10	3.5	8	10			
* Racehorse cubes	13	6.5	6	13			
Forages							
Good grass hay	10	1.6	32	10	0.4	0.25	85
Av grass hay	8	1.6	33–37	9	0.4	0.2	85
Poor grass hay	4	1.6	37+	8	0.3	0.2	85
Silage	13–17	4.0	30	12			25
Haylage	16	2.5	30	11–12			40
* Hydroponic grass	16	4.0	12–15	17			19

* Manufacturer's data.

(3) *Intermediate feeds*. These include wheatbran, sugar beet pulp and
grass meal.
(4) *Forages*. Grass, hay, haylage and silage.
(5) *Compound feeds*. Nuts and coarse mixes.

These feeds will now be considered in more depth, and reference can be
made to the table of nutritive values (table 3.1).

Cereals

Oats

Oats have a high energy value due to their starch content, and are the
traditional grain fed to working horses in this country and northern
Europe. If – and only if – they are of good quality, oats can provide
enough protein for the horse in hard work.

The grain should be plump, with more kernel in relation to husk, shiny
and dust-free, and a pale yellow colour with a sweet smell. Black oats are
also available but are becoming increasingly uncommon.

Oats are fed bruised; the husk should be broken without damaging the
kernel. Once bruised, they should be used within three weeks as the
nutritive value gradually deteriorates. Boiling increases digestibility and is
useful for a horse in poor condition.

Oats can make up 90% of the concentrate ration, but it must be
remembered that they are low in the amino acids lysine and methionine,
with a poor calcium to phosphorus ratio.

Barley

Barley has a higher energy content than oats – hence its fattening
properties – but a lower fibre content, which results in a lower maximum
inclusion rate in the ration of 50%.

The grain should be full, clean and shiny, with a rounder, plumper
appearance than oats. There should be no sharp awns amongst the grains.

Barley can be fed rolled or boiled or heat-treated (micronised). These
treatments all increase digestibility and palatability. Rolled barley can be
crumbly and dusty, and the micronised form, although more expensive,
is becoming increasingly popular and is included at high levels in some
coarse mixes.

Barley seems less likely to make horses overexcited or 'gassy' than oats,
but this property is hard to explain scientifically. Like oats, barley has a
poor calcium to phosphorus ratio.

Maize

Maize has a high energy content and contains little fibre; it is rarely included at more than 25% of the concentrate ration. It appears to be quite 'heating', and too much maize may produce flat lumps under the skin. Maize is generally fed micronised and flaked, and should be bright golden in colour, crisp and clean.

Protein Feeds

Beans and Peas

These can be fed crushed, split or micronised. Peas and locust beans, which are brown and sweet-tasting, are frequently found in coarse mixes. Not only do they have an energy content equivalent to that of oats, but they also have a high protein value. They are suitable for horses in hard work or out-wintered stock, but they appear to be very 'heating' and should be fed with discretion.

Linseed

Linseed is the seed of the flax plant and is high in protein and fat, hence its high energy content although its protein quality is not high. The seed should be small, flat, shiny and dark brown, and it has to be carefully prepared: the linseed should be soaked overnight (for at least six hours) and then brought to the boil and simmered for at least one hour, or until the seeds have ruptured and the liquid is thick and jelly-like. This liquid can be added to the feed cold, or to bran hot to make an appetising bran mash.

Linseed can now be cooked conveniently in microwave ovens, thus overcoming cooking problems.

Intermediate Feeds

Bran

Bran is fed as a filler and as a binder for feed, or as a palatable, slightly laxative mash. The energy content is low, thus bran is not 'heating' and is useful for feeding resting horses. Although the protein content is high, it is not good-quality protein.

Bran is the inner husk of the wheat grain and should ideally consist of broad, pinkish sweet-smelling flakes. Good bran is hard to come by, and rather than feed expensive, poor quality bran it may be better to use hay chaff or sugar beet pulp to bulk out feeds.

Bran affects calcium uptake, and a calcium supplement should be fed when bran is used.

Sugar Beet Pulp

Molassed sugar beet pulp is a useful source of energy and digestible fibre, energy being instantly available from the sugar which is easily digested. Energy is also supplied during fermentation of the digestible fibre in the large intestine. This is in contrast to the fibre in, for example, bran, which is largely indigestible and merely acts as a filler.

Sugar beet pulp is palatable and can be included at levels of up to 10% (dry weight) of the corn ration, even in horses doing hard work. It must be soaked in its own volume of water overnight.

The high calcium to phosphorus ratio in sugar beet pulp can help right the imbalance caused by cereals in the ration.

Forages

The horse's natural food is grass. Grass can be conserved for winter feeding in several ways – hay, silage and haylage, are all used to provide bulk in the horse's ration. Bulk must make up at least 25% of the horse's total food intake, an absolute minimum being 0.7 kg per 100 kg body weight, i.e. 3.5 kg (8 lb) hay per day for a 500 kg horse in fast work. For a horse in light work, bulk would probably make up about 75% of the ration.

Conserved grass can only be as good as the original grass minus conservation losses; so when assessing such food consider the crop from which it came, then consider how well it has been conserved.

Hay

The quality of hay feed can make or break a diet. Good hay should be greenish, smell sweet, and be dust and mould free. (Barn-dried hay has less smell; it also has a higher energy and protein content.) Dusty hay can damage a horse's wind, and should be thoroughly damped if there is no alternative to feeding it. Mouldy hay should never be fed.

Since 80% of the nutrients of grass are in the leaf, stemmy hay offers a lower feed value; it is also less digestible. Hay should be cut when the grass is in flower, but is often cut later than this to gain a heavier yield. If it is cut too late, the seed heads ripen and are shed; thus the feed value of the grass, and hence the hay, falls. This can lead to the incongruous situation where early-cut hay, even if not very well made, can be of better feed value than late-cut more attractive-looking hay.

Buying good-quality hay, although it is expensive initially, pays off in the long run, with horses requiring fewer concentrates and staying healthier. As can be seen from table 3.1 while the energy content of good and bad-quality hay varies little, the protein content can vary dramatically. This can be very important for horses in hard work, and the concentrate ration will have to make up this shortfall.

There are three common types of hay. *Meadow hay* is made from permanent pasture containing vetches and herbs, and is usually soft hay with a low protein content. *Seeds hay* is made from rye-grass-based leys, giving a coarser, harder hay with higher protein values. (A ley is grass sown as a crop and left down for from one to eight years.) Due to the coarser grass types, seeds hay needs a long spell of good weather to be made well. Otherwise it has to be conditioned, or lacerated by machine, to enable it to dry more quickly, but if it is rained on after this, much of the nutritive value is washed out of the hay. *Lucerne (alfalfa)* is a legume grown in small quantities in the east of England. Due to the price and high protein value, its main use is to supplement grass-hay ration. Good lucerne hay is usually barn dried. Sainfoin hay is similar but less common.

Silage

The best silage for horses is grass pickled in its own juice. However, additives are often used and the horse owner should be aware of this. Grass used for silage is cut earlier than for hay – at or just after heading; thus it tends to be a slightly richer feed. The grass is placed in a silo (clamp, tower or heavy-gauge plastic bag). It may be wilted by allowing it to dry slightly before it is ensiled.

In all silage-making processes, as much air as possible is removed and an air-tight seal made. The micro-organisms present in the grass cause fermentation, and acids are produced which prevent the growth of further, destructive micro-organisms and preserve the grass in much the same way that acetic acid (vinegar) preserves pickles. Horses have been successfully fed silage for many years, and the Irish National Stud has carried out successful feeding trials. Once horses have been gradually introduced to silage, they often eat silage in preference to hay. Recently, there have been problems involving contamination of improperly made big-bale silage, and until more is known about feeding silage it may be wise to seek advice before feeding the big-bale type. Silage, like hay, must be free from mould and other contaminents. The main problem in feeding silage to horses is knowing how to handle it.

Haylage

The baled form is usually fed to horses, and it is best described as a compromise between hay and silage: the grass is left to grow slightly longer than it is for silage and is cut between heading and flowering. The grass is left to partially dry, and is then baled and vacuum-packed in tough plastic bags, resulting in dust-free, palatable forage with a high energy and protein content. It is an expensive form of forage to feed and has a high water content; it does not keep well and must be used within three days of opening. However, it is excellent for horses with wind problems. Care must be taken to avoid boredom problems, because a horse can eat its day's ration very quickly.

It is generally accepted that silage and haylage have a higher feeding value than hay. This is mainly due to three factors:

(1) Hay is cut at a later stage of its growth when the grass contains more indigestible fibre.
(2) Hay is very vulnerable to rain while it is drying on the field. Nutrients will be leached from the grass, and the resulting forage will therefore be of poorer quality. Haylage and silage are conserved immediately, with minimum loss of nutrients.
(3) A lot of hay is made from less nutritious species of grass, while silage and haylage are made from high-protein, high-carbohydrate rye grass species.

Hydroponic 'Grass'

Hydroponics is a method of growing forage in water, without soil, in specially heated, lit and irrigated machines (figs 3.4 and 3.5). The 'grass' is grown from soaked barley seeds and takes 5 to 14 days to grow from 'planting' to harvest. Thus fresh, green grass is available 365 days of the year. The nutrient value is similar to that of spring grass, and while the low dry matter makes it impractical as the only source of forage for the eventer it is very useful to maintain appetite as part of the forage ration.

Compound Feeds

These are feeds which have been formulated and mixed together to form a balanced ration. There are four main types of compounds available:

(1) Complete cubes: to be fed alone, and contain both the forage and the concentrate part of the ration.
(2) Concentrate cubes: formulated as complete concentrate rations to

Fig. 3.4 A small hydroponic unit

be fed with normal roughage.

(3) Protein concentrate: high-protein pellets designed for dilution with cereals and other constituents to form a balanced ration which is fed with roughage.

(4) Coarse mixes of cereals, intermediate feeds and other nutrients.

Compound feeds contain a mineral and vitamin premix, and the amount and type of supplement fed will be affected according to the amount of compound fed. If half the food is made up of a cube, the chosen supplement may only have to be fed at half the recommended dose. This rule of thumb will vary depending on the type of cube and type of supplement used. Since horse and pony cubes generally have a low mineral and vitamin level, the competition horse may require additional supplementation.

Coarse mixes range from mixtures of horse and pony cubes with rolled oats, barley and maize, to mixes of high-protein cubes, peas, beans,

Fig. 3.5 Hydroponic grass being fed

linseed cake, locust bean, etc. These mixes are usually very palatable to the horse and attractive to the owner – which is equally important! This is due to the mixture of ingredients being bound together by corn syrup or molasses. Coarse mixes are designed to be fed with hay alone, and it is pointless to mix cereals with them.

Feeding cubes has several advantages, including economy of labour, convenience of feeding, and knowing that a balanced ration of known quality is being fed. The analysis of cubes and coarse mixes has to be declared and must be within certain legally defined limits, unlike cereals which can vary dramatically between batches. Hopefully, the information on the bag will one day be in a more useful form than it is at present.

Compound feeds cannot be correctly balanced for every horse all of the time. A horse may require more energy than is provided by horse and pony cubes, and yet may hot up on racehorse cubes. Provided the feed values are known, different types of cube can be blended to produce a suitable ration. There are many different types of compound feed available, one or more of which may suit a particular horse. It must always be remembered that horses have individual preferences, and the palatability of cubes must be taken into consideration.

The rules of feeding, the nutrient reequirements of horses, and the nutrient content of feedstuffs can be brought together to plan simple rations for horses competing and working at different levels.

The Theory of Rationing

A 'ration' is an allocation of chosen foodstuffs sufficient to provide the daily nutrient requirements of a particular horse. It can be calculated very easily by any horseman provided he follows a few simple rules and has access to a table giving the nutritive value of the feedstuffs available.

Horses are fed for two main reasons: firstly to keep them alive, and secondly so that they can perform the tasks we ask of them. In scientific terms, we call this 'maintenance' and 'production'.

Maintenance

Feeding a horse for maintenance is giving it enough food to maintain it in its present state. This means providing energy for the muscles of the gut, heart and lungs so that essential processes can take place, energy for grazing, for maintaining body temperature and for replacing cells to keep the body in good working order. The hunter turned away during the summer is being fed for maintenance and survives happily on grass alone.

The maintenance requirements of a horse can usually be provided for by forage alone. The major factor which determines how much energy a horse needs for maintenance is its bodyweight – bigger horses need more food just to stay alive.

Production

Production may be divided into six different forms: growth, pregnancy, lactation, fattening, work and repair (or recovery from illness and injury). In this book we are concerned with work, which will fall into various categories depending on the age and ability of horse and rider.

The extra energy and protein required by the horse for work are usually provided by concentrates. The competition horse would not be able to extract enough energy or protein from a forage ration to perform the work without losing condition.

The energy content of a foodstuff is measured in megajoules. Thus the energy content of oats is described as 14 MJ of DE/kg – i.e. each kilogram contains 14 megajoules of digestible energy (see table 3.1). Megajoules are merely metric calories; all we are doing is 'calorie-counting' for our horse.

Protein is measured as a percentage. Oats contain 11% crude protein, each kilogram (1000 g) of oats containing 110 g (11%) of crude protein (see table 3.1).

The Rules of Rationing

There are five steps involved in ration calculation, followed by checking and adjusting the ration to suit individual horses. Each step will be explained and a ration actually calculated afterwards.

Step one: *Estimation of bodyweight*

Several methods are available:

- table of weights (see table 3.2)
- calculation
- weightape
- weighbridge

Table 3.2 Approximate bodyweights

Type	Height (hands)	Approx. weight kg	lb
Pony	13.0	260	570
Large pony	14.2	400	880
Small hunter	15.2	450	985
Medium hunter	16.0	550	1190
Heavy hunter	16.3	600	1410
Showjumper	16.0	500	1100
Shire	17+	1070	2350

Calculation
The approximate bodyweight can be calculated by taking two measurements: one around the girth (G) and one from the point of the shoulder to the point of the hip (L).

$$\text{bodyweight (kg)} = \frac{G \text{ (cm)} \times G \times L \text{ (cm)}}{8700}$$

$$\text{bodyweight (lb)} = \frac{G \text{ (inches)} \times G \times L \text{ (inches)}}{241.3}$$

Weightape
Tapes can be bought which when placed round the girth will show approximate bodyweight.

Weighbridge

This is the only accurate way of assessing a horse's weight. A nearby racing yard may have a weighbridge which they will let you use. Otherwise take your horse to the nearest public weighbridge – and take an old carpet, too, as horses do not like walking on sheet metal.

Step two: *The horse's appetite*

It is futile to place more food in front of the horse than it is physically capable of eating; the ration must be within the appetite of the horse. Appetite is governed by bodyweight:

$$\text{Appetite (kg)} = \frac{\text{bodyweight} \times 2.5}{100}$$

Step three: *Providing energy for maintenance*

$$\begin{array}{l} \text{Energy required for} \\ \text{maintenance (MJDE/day)} \end{array} = 18 + \frac{\text{bodyweight (kg)}}{10}$$

Step four: *Providing energy for work*

For each 50 kg of bodyweight, add the following work score to the maintenance requirement:

One hour walking	+ 1 MJDE
Walking and trotting	+ 2 MJDE
Some cantering	+ 3 MJDE
Schooling, dressage and jumping	+ 4 MJDE
Novice O.D.E. Hunting 1 day/wk	+ 5 MJDE
Hunting 2 days/wk	+ 6 MJDE
Three-day eventing	+ 7 MJDE
Racing	+ 8 MJDE

You will, by adding together the energy requirements for maintenance and work, have the *total energy requirement* for your horse. You can now divide this energy between hay and concentrates, depending on the type of work the horse is doing.

	Energy from hay (%)	Energy from concentrates (%)
Maintenance	100	0
Light work (work score 1–2)	70	30
Medium work (3–4)	50	50
Hard work (5–6)	30	70
Fast work (7–8)	25	75

Step five: *Providing protein for work*

Energy is the first limiting factor in the diet, and if high quality feedstuffs are being fed the protein requirements are likely to be satisfied.

	Crude protein in the ration (%)
Light work	7.5–8.5
Medium work	7.5–8.5
Hard work	9.5–10
Fast work	9.5–10

Checking and Adjusting the Ration

All horses are individuals and must be fed accordingly. Once a ration has been calculated and is being fed, the horse must be monitored to ensure that the ration is suitable.

(1) A supply of clean, wholesome water must be available to the horse at all times.

(2) The foods must be of good quality and acceptable to the horse – is the horse enjoying its food?

(3) Not only must the foods satisfy the horse's nutritional require-ments; the horse must also be psychologically satisfied, i.e. it must not suffer boredom or have a craving for roughage.

(4) The horse's condition must be checked by eye, tape or weigh-bridge. The horse may be gaining or losing weight: is this satisfactory? If not, alter the ration accordingly. Horses have optimum performance weights, and the horseman should be aware of this weight for each individual.

(5) The horse's temperament and behaviour may affect the ration fed. Part-bred horses may need more corn and less bulk as they are better doers and (usually) more placid. Routine and a quiet yard may save feed as horses are not fretting in their boxes.

(6) The horse's environment must be noticed. In a cold spell, more food and an extra blanket may be needed.

(7) Horses must be regularly wormed and have their teeth checked for sharp edges.

(8) Some are poor doers, perhaps due to a gut damaged by worms early in life, and will always need extra attention in their feeding.

Ration Calculation

Example one: A novice one-day eventer, 16.0 hh, Thoroughbred.

Step one: *Estimation of bodyweight*

By table: 500 kg

Using tables of bodyweight, while not ideal, is adequate for this example.

Step two: *Appetite*

$$\text{Appetite (kg)} = \frac{\text{bodyweight kg} \times 2.5}{100}$$
$$= \frac{500 \times 2.5}{100}$$
$$= 12.5 \text{ kg } (27.5 \text{ lb})$$

Step three: *Providing energy for maintenance*

$$\text{Energy for maintenance (MJDE)} = 18 + \frac{\text{bodyweight (kg)}}{10}$$
$$= 18 + \frac{500}{10}$$

The maintenance requirement is 68 MJDE.

Step four: *Providing energy for work*

Novice one-day event: + 4 MJDE per 50 kg bodyweight

$$\text{Energy for work} = 4 \times \frac{500}{50}$$

The work requirement is 40 MJDE.

Total energy requirement = 68 + 40 MJDE
$$= 108 \text{ MJDE}$$

From the table, for a horse in medium work, provide 50% of the energy from hay and 50% concentrates.

Energy from hay $= 0.50 \times 108$
$$= 54 \text{ MJDE}$$
Energy from concentrates $= 0.5 \times 108$
$$= 54 \text{ MJDE}$$

Using the table of nutritive values:
We must provide 54 MJDE from hay. Average grass hay contained 9 MJDE/kg.

Hay fed (kg) = $\frac{54}{9}$ = 6 kg

Therefore 6 kg (about 14 lb) of hay must be fed every day.

We must provide 54 MJDE from concentrates. Oats contain 14 MJDE/kg.

Oats fed (kg) = $\frac{54}{14}$ = 4 kg

Therefore 4 kg (nearly 9 lb) of oats must be fed every day.

The energy requirements of a novice one-day event horse would be satisfied by a ration of 6 kg of hay and 4 kg of oats (well within appetite limits). If you do not want to feed solely oats, just play with the figures:

2 kg oats	@	14 MJDE/kg	=	28
0.5 kg beet pulp	@	14 MJDE/kg	=	7
2 kg horse and pony cubes	@	10 MJDE/kg	=	20
4.5 kg				55 MJDE

This ration is practical and supplies adequate energy.

Step Five: *Providing protein for work*

A horse in medium work requires 7.5 to 8.5% crude protein. The protein content of the ration can be worked out using the table of nutritive values.

The percentage protein in
the final ration $= \frac{95}{10.5}$

$$= 9\%$$

			Protein content of feed (%)		Protein in ration (g)
Hay	6 kg	×	8	=	48
Oats	2 kg	×	11	=	22
Sugar pulp	0.5 kg	×	10	=	5
Cubes	2 kg	×	10	=	20
	10.5 kg				95.0

The ration provides more than adequate protein.

Example two:
A three-day event horse, 16.2 hh, Thoroughbred.

Step one: *Estimation of body weight*

By table: 550 kg

Step two: *Appetite*

Appetite (kg) = $\dfrac{\text{bodyweight (kg)} \times 2.5}{100}$

= $\dfrac{550 \times 2.5}{100}$

Appetite = 13.75 kg (30.25 lb)

Step three: *Providing energy for maintenance*

Energy for maintenance (MJDE) = $18 + \dfrac{\text{bodyweight (kg)}}{10}$

= $18 + \dfrac{550}{10}$

The maintenance requirement is 73 MJDE.

Step four: *Providing energy for work*

Three-day eventing: + 7 MJDE per 50 kg bodyweight

Energy for work $= 7 \times \dfrac{550}{5}$

The work requirement is 77 MJDE.

Total energy requirement $= 73 + 77$ MJDE
$$= 150 \text{ MJDE}$$

From the table, for a horse in fast work provide 25% of the energy from hay and 75% from concentrates.

Energy from hay $= 0.25 \times 150$
$$= 38 \text{ MJDE}$$
Energy from concentrates $= 0.75 \times 150$
$$= 112 \text{ MJDE}$$

Using the table of nutritive values:
We must provide 38 MJDE from hay. Average grass hay contains 9 MJDE/kg.

Hay fed (kg) $= \dfrac{38}{9} = 4 \text{ kg}$

Therefore 4 kg (nearly 9 lb) of hay must be fed every day.

An absolute minimum of forage is 0.7 kg per 100 kg bodyweight, i.e. 3.9 kg (8.5 lb) for our 550 kg eventer.

We must provide 112 MJDE from concentrates. Racehorse cubes contain 13 MJDE/kg.

Mix fed (kg) $= \dfrac{112}{13} = 8.5 \text{ kg}$

Therefore 8.5 kg (about 18 lb) of mix must be fed every day.

The energy requirements of a three-day event horse would be satisfied by a ration of 4 kg hay and 8.5 kg of coarse mix (well within the appetite level of nearly 14 kg).

An alternative ration would be:

Sugar beet pulp	1 kg	@	14 MJDE/kg	14
Oats	4 kg	@	14 MJDE/kg	56
Barley	1 kg	@	15 MJDE/kg	15
Maize	1 kg	@	17 MJDE/kg	17
Grass nuts	0.75 kg	@	13 MJDE/kg	10
Linseed	0.25 kg	@	27 MJDE/kg	7
	8 kg			119 MJDE

Step five: *Providing protein for work*

A horse in fast work requires 9.5 to 10.5% crude protein in the ration. The protein content is worked out as before.

			Protein content of food (%)		Protein in ration (g)
Hay	4 kg	×	8	=	32
Sugar beet pulp	1 kg	×	10	=	10
Oats	4 kg	×	11	=	44
Barley	1 kg	×	11	=	11
Maize	1 kg	×	10	=	10
Grass nuts	0.75 kg	×	17	=	13
Linseed	0.25 kg	×	26	=	6
	12 kg				126

The percentage protein
in the final ration

$$= \frac{126}{12}$$

$$= 10.5\%$$

The ration provides adequate protein.

In conclusion, these rations are only specimens of the type of calculation which can be used. These calculations are not intended to take the art out of feeding horses, but to remove some of the guesswork. The 'art' comes in keeping horses happy and in peak condition for long periods of time and recognising which ration suits an individual horse. There are no formulas for this – just experience, for which there is no substitute.

4 The Respiratory System

Respiration is the process by which the horse takes oxygen into its body and also rids its body of carbon dioxide. This process takes place in every part of the body, the cells in the living tissue taking oxygen from the blood and giving back carbon dioxide.

Anatomy

The respiratory system consists of the airways of the head and neck and the lungs.

Within the Head and Neck

The airways in the head and neck are shown in fig. 4.1. The *nostrils* are the entrance to the respiratory system; the horse draws air in only through the nostrils and not through the mouth, and they can expand in order to take in more air. The nostrils lead to the *nasal passages*, where the air is warmed and foreign particles, e.g. dirt, are screened. Air then passes to the *pharynx* at the back of the throat. At the bottom of the pharynx is the *soft palate* which is overlapped by the *epiglottis*.

Most of the time this overlapping shuts off the mouth and allows the free passage of air into the next part of the system, the *larynx*. When the horse swallows, however, the epiglottis flips over the opening of the larynx, the soft palate moves up and food is admitted from the mouth into the pharynx and is then swallowed (fig. 4.2). This mechanism is effective in separating the processes of breathing and eating but makes mouth-breathing under stress very difficult for the horse.

The air then passes to the larynx, which is a hollow box sitting on top of the *trachea* (or windpipe). It contains the *vocal cords* and controls the passage of air in and out of the trachea. The trachea runs from the larynx to the lungs, and consists of a large tube held open by rings of cartilage.

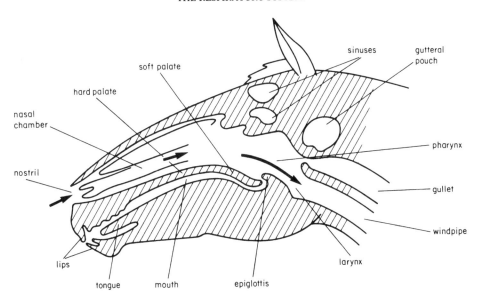

Fig. 4.1 The airways of the head and neck

Within the Chest

The structure of the lungs is shown in fig. 4.3. The trachea divides into two *bronchi* close to the lungs, each *bronchus* supplying one lung. Within the lung they divide and subdivide to give small *bronchioles*. The bronchi are supported by rings of cartilage but the bronchioles are not. The bronchioles end in tiny sacs, which under the microscope look like bunches of grapes and are called *alveoli*. These sacs make up the body of the lung and make it feel like a smooth sponge to the touch.

There are thousands of millions of alveoli, each about 0.3 mm across, giving a total surface area of several hundred square metres. Each alveolus is wrapped around by tiny blood vessels called *capillaries* (fig. 4.4) and blood flows from the capillaries into the thin membranes of the alveoli. The air contained within the sacs is in such close contact with the blood that oxygen is able to move from the air in the alveoli into the blood, and carbon dioxide from the blood diffuses into the air in the alveoli and is removed when the horse breathes out.

The *diaphragm* (or midriff) separates the chest and lungs from the abdomen containing the digestive organs. It is a strong, dome-shaped sheet of muscle attached to the ribs and is used in breathing.

Figure 4.5 shows how the lungs are arranged in the thoracic cavity (chest) of the horse in relation to the heart.

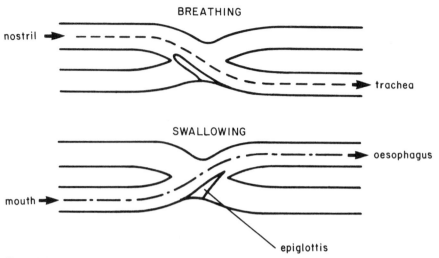

Fig. 4.2 The operation of the epiglottis in the pharynx during breathing and swallowing

Respiration

The process of respiration can be broadly divided into two parts:

(1) Breathing: this is the obvious and well known part of the process.
(2) Tissue respiration: this is the aspect which is so important to the trainer of competition horses.

Breathing

Breathing is the process of taking air down into the lungs and then expelling it from the lungs.

The horse draws air in through its nostrils; air passes through the larynx, down the trachea and into the lungs. Here, oxygen passes from the air into the blood and carbon dioxide passes from the blood into the lungs, a process called gaseous exchange. The lungs are filled by the action of the dome-shaped diaphragm and the ribs. When the ribs are pulled outward, the chest expands; simultaneously the diaphragm contracts, flattening the dome thus enlarging the 'box' in which the lungs are contained and drawing air in to fill the available space. This process is called inhalation.

Breathing out, or exhalation is achieved by the elastic recoil of the rib cage and lungs so that air is forced out. This may also be assisted by contraction of the abdominal muscles pushing the gut contents against the diaphragm. Thus air in the lungs is exchanged with air from outside and oxygen is brought to the alveoli. Oxygen and carbon dioxide are exchanged by a process of simple diffusion – that is, by moving from the area of high to low pressure on each side of the thin alveolar membrane that separates air from the blood in the capillaries (see fig. 4.4).

Without oxygen, the muscles cannot function. Maximum performance of the horse depends on the effective passage of oxygen to the muscles and other tissues of the body. Oxygen is carried in the blood by means of the respiratory pigment, *haemoglobin*, in the red blood cells; this will be discussed further when dealing with the circulatory system.

At rest, the horse has a respiration rate of 8 to 16 breaths per minute, the rate being higher for youngstock. After strenuous exercise, this can increase dramatically to 120 breaths per minute. Table 4.1 shows the resting respiratory rates for various other domestic animals and man.

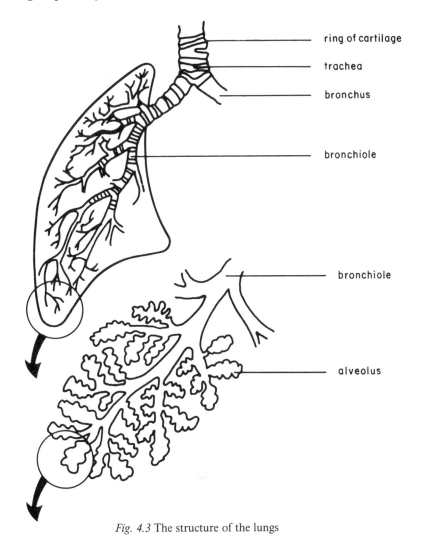

ring of cartilage

trachea

bronchus

bronchiole

bronchiole

alveolus

Fig. 4.3 The structure of the lungs

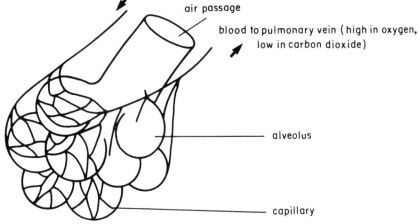

blood from pulmonary artery (low in oxygen and high in carbon dioxide)

air passage

blood to pulmonary vein (high in oxygen, low in carbon dioxide)

alveolus

capillary

Fig. 4.4 The alveoli, showing blood supply

Table 4.1   Resting respiratory rates

Rate per minute	
Horse	8–12
Cow	15–24
Sheep	12–24
Goat	12–20
Pig	15–24
Dog	19–30
Cat	24–42
Man	12–30

Tissue Respiration

The second part of the respiratory process takes place when the oxygen being carried in the blood reaches the muscle tissue, and is called tissue respiration. This process produces energy from glucose in the blood or from fuel stored in the muscle cells as glycogen. To release the energy from the fuel supply requires oxygen which is carried in the blood; the process involves a complex series of reactions.

Energy can be created by two methods of tissue respiration:

(1) Aerobic respiration.
(2) Anaerobic respiration.

Aerobic (with oxygen) respiration takes place when the oxygen supply arriving in the blood is adequate to 'burn up' some of the glycogen in the muscle and to supply the energy demands of the animal. During this process, which is known as 'glycolysis', 1 unit of oxygen produces 36 units of energy. Carbon dioxide is also produced as a waste product; this is eventually expelled from the lungs during exhalation (fig. 4.6).

Anaerobic (without oxygen) respiration takes place when the energy

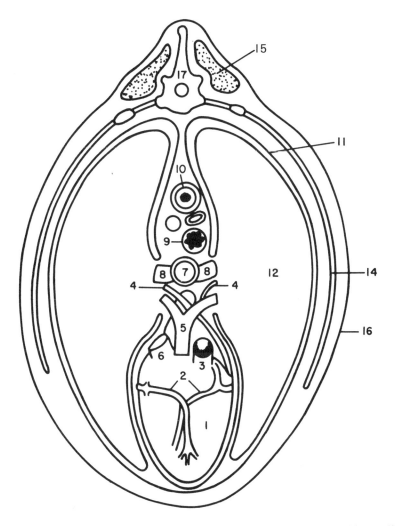

Fig. 4.5 The thoracic cavity, showing the relative positions of the lungs and heart: (1) Heart (2) Coronary Vessels (3) Aorta (4) Pulmonary Veins (5) Pulmonary Arteries (6) Vena Cava (7) Trachea/Windpipe (8) Bronchi (9) Oesophagus/Gullet (10) Aorta (11) Pleural Membrane (12) Lungs (13) Vertebra (14) Ribs (15) Back muscles (16) Skin

demands of the animal are greater than can be suplied by the oxygen arriving in the blood. This situation is found in maximal exertion, e.g. sprinting. The muscle tissue then uses glycogen reserves to produce energy without the aid of oxygen. However, 1 unit of glycogen only gives 3 units of energy, thus anaerobic respiration is only one-twelfth as

(a) Aerobic metabolism

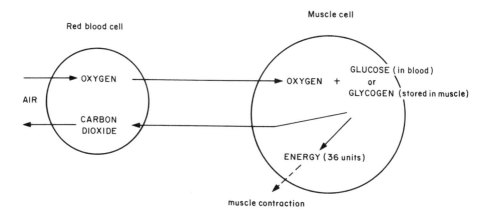

Aerobic metabolism is a slow process, the rate is limited by oxygen transport in the blood

(b) Anaerobic metabolism

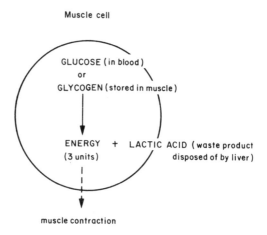

Fig. 4.6 Energy sources for work

efficient as aerobic respiration (see fig. 4.6).

Anaerobic respiration causes a poisonous waste product called lactic acid to be produced. Lactic acid is removed in the blood stream and broken down by the liver into harmless substances, a process called detoxification. However, if the lactic acid is being produced in large amounts, as in the case of maximal exertion, e.g. galloping, lactic acid accumulates because the blood cannot remove it quickly enough. This build-up of lactic acid poisons the muscle cells and they can no longer contract. In practical terms, a horse cannot gallop indefinitely; tiredness begins to set in. Lactic acid contributes to fatigue and is a major limitation to athletic performance.

It follows that the longer a horse can work aerobically, with no lactic acid production, before anaerobic respiration and lactic acid build-up occur, the longer it can carry on before fatigue stops work. This decribes, in physiological terms, the basic aim of getting a horse fit, and the words 'aerobic' and 'anaerobic' will be used frequently in the course of the book.

The Effects of Training

Both rider and trainer are strongly aware of the necessity of knowing that a horse is 'right in its wind', and much of a training programme is devoted to getting a horse fit and clear in its lungs. A fit horse is able to do a lot more work than an unfit horse before it starts to blow. Once blowing, a fit horse's breathing rate will return to normal more quickly than an unfit horse's.

Training accomplishes the following changes in the respiratory system:

(1) *Alveolar recruitment.* During periods of inactivity, the tiny air sacs in the lungs become blocked by mucus and other debris. Exercise causes this debris to be brought up out of the lungs. This is the reason for an unfit horse occasionally being 'thick in its wind' and having a slight discharge of mucus from its nostrils when it is first brought into work. This clearing of a horse's wind is called alveolar recruitment, and results in the horse having a larger useful lung capacity and hence a greater ability to get oxygen into the blood.

(2) *Pulmonary capillarisation.* The tiny blood vessels or capillaries that surround each alveolus respond to the increased oxygen demands of the exercising horse by increasing in number, or proliferating. Thus there is not only a greater area of blood vessels for gaseous exchange to take place across, but also a greater area of lung tissue and hence more oxygen arriving at the tissues.

(3) *Muscular development.* The muscles of the diaphragm and chest

which control breathing may also get stronger as the exercising horse uses them more.

(4) *Nostril size.* There is a school of thought that believes that nostril size may be limiting to air intake. Horses bred for performance, e.g. Arabs and Thoroughbreds, have nostrils which are capable of flaring widely when the horse is worked. Certainly any restriction in the airways will inhibit performance.

Despite these effects, a horse's wind must be basically sound if it is to be trained to maximum fitness. A defect such as roaring or whistling, or an allergy, may affect the wind in such a way that the afflicted horse will never be able to reach peak fitness. This does not imply that such horses are not capable of hunting or eventing, but it will not be possible to get them as fit as if there were no defect.

The object of the horse's training programme is to teach the body to use more oxygen. The more oxygen the body can use, the more work can be done aerobically: the 'aerobic threshold' of the animal has been increased. Thus the onset of anaerobic respiration is delayed, and hence the production of lactic acid and the onset of fatigue are postponed. Anaerobic respiration is a 'spend now, pay later' process: the energy supplies used have to be replaced, and this process requires oxygen. In effect, an 'oxygen debt' has been created, which has to be paid off after strenuous exercise – this is why the horse goes on blowing after it has pulled up. Thus the horse's return to normal breathing after strong exercise can give a measure for fitness.

The energy produced during tissue respiration may not be needed immediately. This energy is stored in a remarkable chemical molecule called adenosine triphosphate (ATP). The stored energy is released when adenosine triphosphate is converted to adenosine diphosphate (ADP) (fig. 4.7).

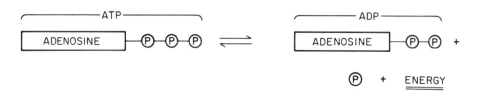

Fig. 4.7 High energy compounds

5 The Muscle System

Muscle does the work that drives the horse, and part of getting a horse fit consists of muscle development or 'muscling-up'. Different shapes and makes of horses develop muscle in varying places: the staying Thoroughbred tends to be lean and rangy, while the sprinting Thoroughbred and the Quarterhorse are blocky and more heavily muscled. In order to understand why these differences arise and how to develop a horse's muscles to maximum fitness, it is important to understand a little of the structure and function of muscle tissue.

In general, muscles are divided into three major groups:

(1) *Smooth muscle* is found in the walls of hollow organs such as the digestive tract.
(2) *Cardiac muscle* is found only in the heart.
(3) *Skeletal or striated muscle* moves the skeleton, and it is this muscle that the rider is concerned with developing.

Skeletal Muscle

Skeletal muscle cells are specialised for the function of contraction. Each skeletal muscle is made up of many millions of elongated cells called *fibres*, or myofibrils. These lie parallel to one another and are bound together and to the tendons at each end of the muscle by connective tissue (see fig. 5.1). Muscle cells contain stores of glycogen and the enzyme systems for the breakdown of the glycogen to give energy for muscle contraction. The cells also contain the red muscle pigment *myoglobin* which, like haemoglobin, can bind oxygen to itself and so acts as an oxygen store.

Muscle contraction is brought about by components of the muscle fibre sliding over one another, leading to a shortening of the muscle (fig. 5.2). This contraction requires energy. The energy supply is maintained by the 'powerhouses' of the cell, tiny structures called *mitochondria*, and the fuel for these powerhouses is obtained from the breakdown of glucose. Energy can also be obtained from the breakdown of free fatty acids, which are present in the blood or stored in the muscle.

Muscles receive nerve messages which stimulate contraction. They also have a continuous blood supply which brings the oxygen and other nutrients necessary for producing the energy for contraction, and takes away carbon dioxide and other waste products. During strenuous activity, the blood supply to muscles can be increased sixtyfold, indicating the importance of the muscles receiving large supplies of nutrients. This also indicates the benefits of training so that each system can operate at maximum efficiency.

Muscle Fibre Types

Over the last ten years, the knowledge of the composition of muscle and how various muscle fibres are used during exercise has increased

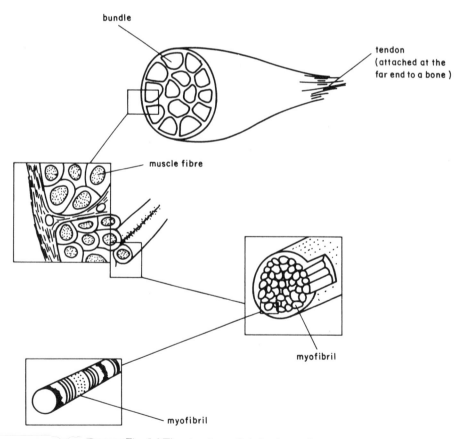

Fig. 5.1 The structure of skeletal muscle

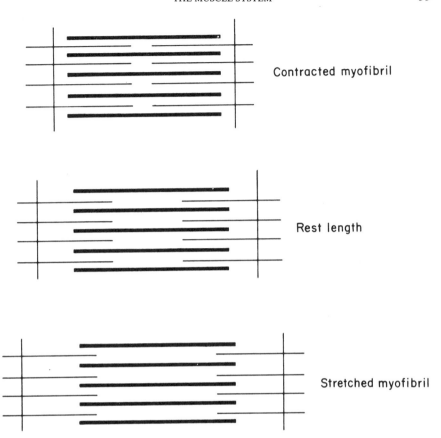

Contracted myofibril

Rest length

Stretched myofibril

Fig. 5.2 Muscle contraction and relaxation

dramatically. This knowledge has been helped by using a muscle biopsy technique, which is a safe painless way of taking tiny samples of living muscle for study.

It has been discovered that the horse possesses different types of skeletal muscle fibres. Initially, muscle types can be identified by colour: red muscle is associated with long-term or endurance work, e.g. leg muscles of fowl. The dark colour reflects the high myoglobin content and thus the greater ability of the muscle to use oxygen, i.e. it is *high oxidative muscle*. White muscle is associated with power, e.g. the breast muscles of fowl, which are the powerful muscles used for flying. This muscle has a lower capacity for using oxygen, i.e. it is *low oxidative muscle*.

More specifically, muscle can be divided into two major types depending on its contractile properties:

(1) *Slow twitch muscle* has a slower contraction time and a greater capacity for using oxygen and can therefore work continuously.

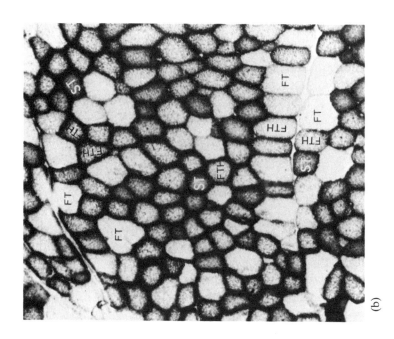

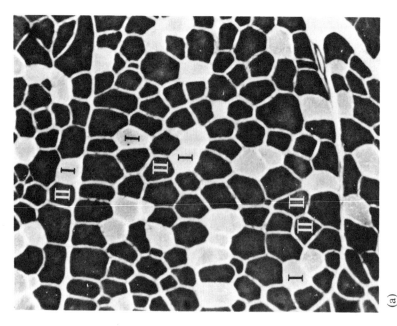

Fig. 5.3 Skeletal muscle fibre types (Snow, 1983): (a) slow (Type I) and fast (Type II) fibres, and (b) Type I high-oxidative fibres (ST), Type II high-oxidative fibres (FTH) and Type II low-oxidative fibres (FT)

(2) *Fast twitch muscle* has a faster contraction time. Two types have been identified according to their ability to use oxygen:

(a) High oxidative, fast contraction muscle, but which is also able to use oxygen and therefore to work for long periods. This makes it very important in the horse required to sustain speed over long distances.

(b) Low oxidative muscle used for galloping and fast acceleration. This muscle fatigues rapidly.

Muscle Fibre Recruitment

During most muscular contractions, it is not necessary for all the muscles to generate their maximum strength. Not all the fibres have to be stimulated. There is an orderly selection of muscle fibres depending on the amount of exertion: for walking and standing, only slow twitch fibres are used; as the speed increases, fast twitch fibres of the high oxidative type are recruited, and it is only during rapid acceleration and jumping that the fast twitch fibres of the low oxidative type are used.

Most muscles contain a mosaic of these three skeletal muscle fibres spread uniformly throughout the entire muscle. The presence of all three fibre types allows a wide range of responses, and the muscle is able to respond to the various demands put upon it (fig. 5.3).

Distribution of Fibre Types

It has been discovered that elite marathon runners have a very high proportion of slow twitch fibres while sprinters have a lower proportion of slow twitch fibres and more fast twitch fibres. This difference appears to be determined genetically, i.e. marathon runners are born with the potential to run long distances.

A similar trend can be seen in different breeds of horse. Even in the untrained state the sprinting specialist, the Quarterhorse, has a greater proportion of fast twitch fibres than the Thoroughbred or Standardbred horse (table 5.1).

Another feature of the Quarterhorse is that the fast twitch muscle fibres have a greater diameter, i.e. they are thicker and are thus able to generate more power. In those horses well adapted for endurance work, all the muscle fibres are of similar size. This allows blood to reach all the fibres, the blood bringing with it adequate supplies of oxygen and nutrients (fig. 5.4).

Fatigue

When getting a horse fit, the aim is to be able to work the horse for longer

Table 5.1 Slow twitch fibres in different breeds of horse (from the middle gluteus muscle)

	%
Quarterhorse	7
Thoroughbred	13
Arab	14
Standardbred	18
Shetland	21
Pony	23
Donkey	24
Endurance horse	28
Heavy horse	31

From Snow (1985).

before it gets tired; in other words, to delay the onset of fatigue. Fatigue is the inability of a physiological function to continue at that level – a horse cannot gallop indefinitely. If the horse is not allowed to slow down, exhaustion will eventually set in – exhaustion being the complete inability to maintain exercise.

It is important to consider the causes of fatigue in order to be able to train a horse and avoid exhaustion. There are four major factors contributing to the onset of tiredness:

- glycogen depletion
- lactic acid build-up
- dehydration and heat stress (hyperthermia)
- lameness

The last two factors will be discussed later.

Glycogen Depletion

During work, the different muscle fibre types are brought into use in turn and progressively depleted of their glycogen stores. A horse performing mainly aerobic (with oxygen) exercise, e.g. prolonged low speed endurance work, will be using a majority of slow twitch fibres and a few fast twitch high oxidative fibres. Oxygen can be brought by the blood to the fibres fast enough to supply the energy demands of the muscle because the horse is not working hard.

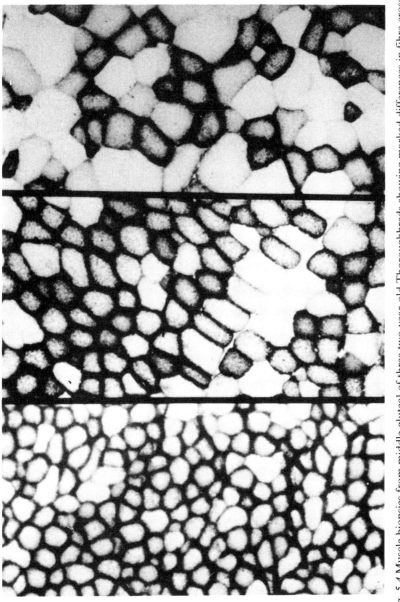

Fig. 5.4 Muscle biopsies from middle gluteal of three two-year-old Thoroughbreds showing marked differences in fibre cross-sectional area (reacted for succinic dehydrogenase activity (Snow, 1983) *The Veterinary Record*)

However, glycogen stores will eventually be used up; there will be no source of energy for the muscle fibres and they will no longer be able to contract. (Remember that oxygen and glycogen are needed to produce energy for contraction during aerobic respiration.) Once all the glycogen has been depleted from muscle cells, it takes 48 to 72 hours to replace it.

Lactic Acid Build-up

As the speed of the work increases, fast twitch low oxidative fibres will be recruited. The blood becomes unable to supply oxygen in large enough amounts for aerobic respiration to take place, and anaerobic (without oxygen) respiration will commence. The by-product of anaerobic respiration, lactic acid, will be produced in the muscles.

At a fast gallop, the lactic acid will be produced at such a rate that the blood cannot carry it away from the muscles quickly enough, resulting in a build-up of lactic acid in the muscles. These acid conditions prevent the muscle fibres from functioning properly. Eventually they cut out and are unable to contract, and the horse has to slow down.

In man, the lactic acid build-up is associated with pain. This could also be true in horses, in which case a horse's performance may also depend on how much pain it feels and how willing it is to withstand that pain. However, the horse is able to tolerate much higher levels of lactic acid before tiring than the human athlete, indicating the horse's adaptation as an elite athlete.

Once the lactic acid has been removed from the muscles, exercise can be continued at similar workloads – think of how the flagging hunter recovers after a check. Removal of the lactic acid is more rapidly accomplished if mild exercise is continued. The racehorse would benefit from being cooled off at a slow trot after a gallop, because of the maintenance of a good blood supply to the muscles to carry the lactic acid away. There may also be long-term advantages in the lactic acid being removed quickly – there may be less stiffness in the muscles next day, which is important in the three-day event horse which has to showjump the day after completing a strenuous cross-country course.

Giving a horse sodium bicarbonate, which is alkaline, to neutralise the acid may prolong the time to fatigue at high workloads. However, no work has been done to discover the optimum amount to give and how long before a race to give it.

Lactic acid, then, is a limiting factor to the performance of the horse; therefore any training programme must aim to delay the build-up of lactic acid.

The Effects of Training

Muscle is able to adapt during training to the varying demands placed upon it. Although it *is* possible to change slow twitch fibres to fast twitch fibres and vice versa, it is unlikely in normal training that there will be any significant changes in the proportions of these two basic types of fibres. The main effects of training are on the metabolism of the fibres, i.e. whether the fast twitch fibres have a high or low oxygen-using capacity. In other words, training can increase the amounts of fast twitch fibres with a high oxygen-using capacity; and the more of these present, the greater the ability of the horse to utilise oxygen.

A muscle fibre can only adapt to the demands made of it in competition if it has received suitable stimuli during training. Thus one can only get a favourable effect from training if the programme includes the same type and similar intensity of exercise as is met during competition. If speed is required, sufficient sprinting should be included to recruit the necessary fibres for power and acceleration. If the competition involves endurance work at varying speeds, prolonged exercise at these levels is needed to increase the oxygen-using capacity of the muscles. This means that the time until the onset of fatigue is delayed, and the horse can work at higher speeds before anaerobic metabolism and the building up of lactic acid begins.

Training increases fuel availability to muscle fibres by encouraging more mitochondria (the powerhouses where energy is generated) and more enzymes (biological catalysts) for the breakdown of glycogen and free fatty acids. The blood supply is also increased as the capillaries feeding the muscle proliferate, and the muscle develops greater ability to store glycogen. Anything that slows down the rate of breakdown of glycogen or increases its concentration in the muscle will prolong the time the horse can work before fatigue sets in. Training also encourages the use of free fatty acids as an energy source. This is known as 'glycogen sparing' as more glycogen is left for emergency use. Human marathon runners are able to manipulate their diet prior to a race so that the concentration of glycogen in the muscles is very high, but use of these techniques in horses is debatable because of the chance of azoturia resulting from high glycogen levels. Also, in the horse the spleen acts as a blood reservoir; there is no equivalent action in man, thus training human and equine athletes are parallel but not identical sciences.

6 The Circulatory System

The circulatory system consists of a muscular pump – the heart – and a network of blood vessels for carrying blood throughout the body.

Functions of the Circulatory System

The basic function of the system is to transport essential substances round the body, including:

(1) *Oxygen* from the lungs to all body cells.
(2) *Carbon dioxide* from body cells to the lungs.
(3) *Nutrients* from the gut to body cells.
(4) *Waste products* from body cells to the kidneys.
(5) *Hormones* from glands to body cells.
(6) *Antibodies*, the body's defence force, to sites of attack.
(7) *Heat*, evenly distributed from the body core to the skin.

Anatomy of the Circulatory System

The Heart

The heart (fig. 6.1) is a hollow muscular pump at the centre of the circulatory system, and consists of four chambers. The upper chambers of the heart are the right and left *atria* which act as receiving chambers for blood coming from large veins. The lower chambers are the right and left *ventricles* which act as pumping chambers, pushing the blood into the large arteries leaving the heart. Due to this pumping action, the ventricles are larger (more muscular) than the artria, the left ventricle being the largest because it pumps blood throughout the body.

Circulation of Blood through Heart and Lungs

The circulation of the blood is illustrated in fig. 6.2. The *vena cava* brings deoxygenated blood back from the body to the right side of the heart.

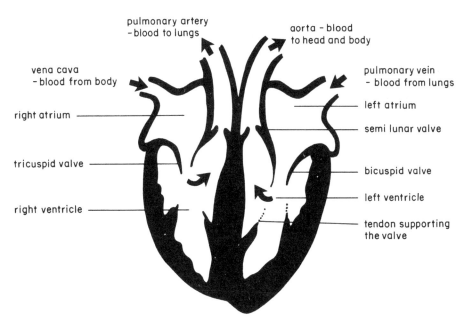

Fig. 6.1 The heart

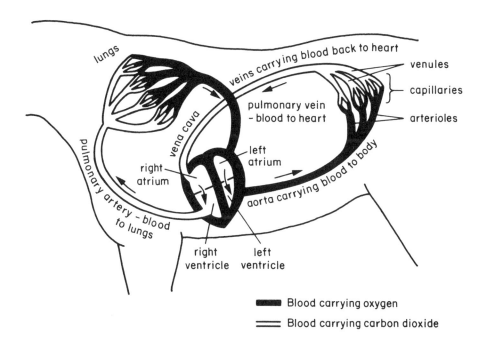

Fig. 6.2 The heart and circulation

Deoxygenated blood has had the majority of its oxygen removed and replaced by carbon dioxide in the tissues (as described in chapter 4). The blood is collected in the right atrium and passes into the right ventricle. The right ventricle contracts and sends blood into the pulmonary artery which leads to the lungs. In the lungs, carbon dioxide is exchanged for oxygen, i.e. the blood is oxygenated and returned to the left side of the lung along the *pulmonary vein*. Blood enters the left atrium, then passes into the left ventricle. When the left ventricle contracts, blood is pushed into the *aorta* (the main artery of the body). Throughout this sequence valves in the heart prevent backflow of blood.

Circulation of Blood through the Body

Oxygenated blood is pumped away from the left ventricle via the aorta and distributed under pressure throughout the body by a branching system of *arteries* and *arterioles* (small arteries). These vessels have thick muscular walls which can control the amount of blood that flows at any time, thus helping maintain blood pressure. The horse owner is aware of arteries as being the places where the pulse is taken, e.g. under the jawbone, under the dock, inside the elbow and below the hock. These are all places where an artery passes over a bone close to the skin and a pulse of blood flowing through the vessel can be felt. This pulse corresponds to the beating of the horse's heart.

As the branches become smaller, the oxygenated blood passes through a system of tiny vessels called the *capillary network*. Here water, oxygen and other nutrients carried in the blood pass through the walls of the capillaries into the body tissue cells, in exchange for waste products from the cells which are carried away in the blood.

The tiny capillaries then merge, forming larger and larger *veins* which carry deoxygenated blood back to the heart. Veins have thin walls with very little muscle. In order to prevent backflow of blood, the walls of veins have a series of one-way valves (figs. 6.3 and 6.4), and this allows muscle contractions from normal body movements to help push blood back to the heart. (The swelling of a horse's legs when exercise is restricted is due to a lack of muscular activity, and hence an inhibition of return of blood to the heart and fluid accumulation in the legs.) The large vein of the body, the vena cava, then returns all the deoxygenated blood to the right side of the heart.

The circulation of blood through the body is called the 'systemic circulation' and that through the lungs the 'pulmonary circulation', and together they make up the characteristic double circulation of the horse.

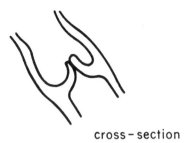

cross – section

Fig. 6.3 Cross-section through a valve

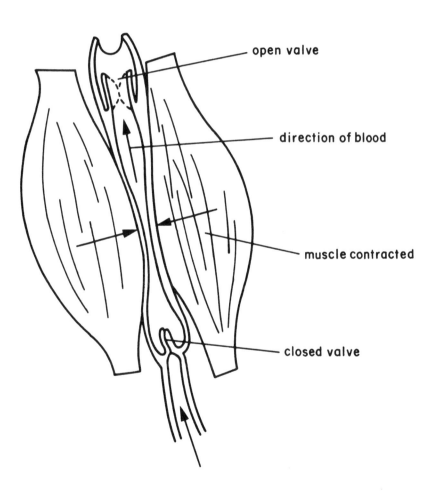

open valve

direction of blood

muscle contracted

closed valve

Fig. 6.4 Valves of a vein, showing the pumping action of adjacent muscles

The Effects of Training and Exercise

The resting heart rate of a horse is normally 36 to 42 beats per minute – that of a fit horse may be as low as 26 beats per minute – and reaches a maximum of 240 per minute when galloping.

As the aim of training is, as has already been stated, to get oxygen to the muscle cells and to remove waste products as efficiently as possible, the heart and circulation have a vital role to play. During exercise, the muscles contract more frequently; their energy requirements (i.e. the oxygen requirement of the body) is greater. Thus the immediate effects of exercise are:

- increased heart rate
- increased respiratory rate

The increased heart rate causes a greater cardiac output – more blood is pumped round the body in a given time. The higher respiratory rate supplies the blood with more oxygen. In an unfit horse, these increases cannot be sustained for long.

The Effects of Training

(1) *Increased heart size.* Like any other muscle in the body, the horse's heart responds to exercise by developing and getting bigger (fig.

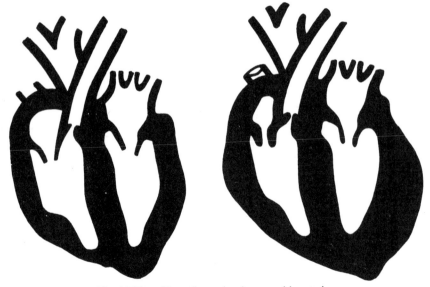

Fig. 6.5 The effect of exercise: increased heart size

6.5). The average size of the horse's heart is about 9 lb; the trained heart can weigh 12 lb and pump three pints of blood at each beat. The heart of the famous racehorse Eclipse weighed over 14 lb. The trained horse galloping sends its blood once round its body in just over five seconds, emphasising the importance of a healthy heart and circulation.

(2) *Capillarisation.* Training increases the number of blood capillaries in the muscles – capillarity can increase by up to 50% (fig. 6.6). This means there is a greater surface area of blood vessels through which oxygen can be supplied to the muscle tissues. If capillaries are not being used, as in the unfit horse, they tend to 'dry up'; training

Training causes capillaries to multiply around the muscle fibres

Fig. 6.6 The effect of exercise: capillarisation of muscles

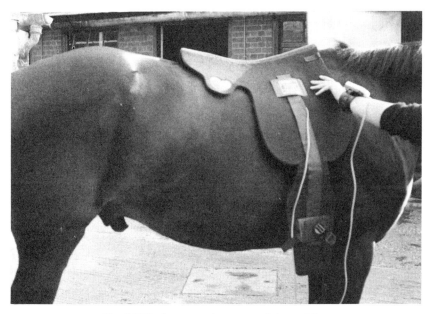

Fig. 6.7 The heart monitor numnah in position

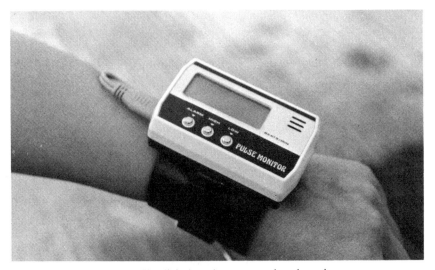

Fig. 6.8 The digital readout strapped to the wrist

causes them gradually to come back into use. It has been estimated that it takes up to three years to realise the full potential of the circulation when bringing a totally unfit horse into work.

These internal changes in the heart, circulation and respiratory system show themselves to the horseman as the horse gets fitter by:

- a slower resting heart rate
- a less dramatic increase in heart rate and respiration rate for a given amount of exercise
- a quicker recovery rate (the time taken for heart rate and/or respiration rate to return from working levels to resting levels)

In simple terms, the horse can work for longer without becoming distressed or fatigued.

Monitoring Heart Rate

Heart rate can be monitored in several ways by the rider/trainer or by the veterinary surgeon:

(1) *Stethoscope.* This can be used by the rider to determine heart rate, or by the vet to detect any abnormalities in heart sounds (this is discussed in the next section).

(2) *Pulse.* As mentioned previously, the pulse is felt where an artery passes over a bone close to the skin, the most commonly used site being the facial artery under the cheek bone. This is a simple way

Fig. 6.9 The monitor in position

for the rider or vet to asses the heart rate of an animal.

(3) *Heart rate monitor.* This is a machine which records the heart rate instantaneously by means of electrodes pressed against the skin (figs. 6.7, 6.8 and 6.9). A digital readout display is attached to the rider's wrist or around the horse's neck so that heart rate can be read at any time during exercise. In the U.S.A., sophisticated versions are considered invaluable for interval training, but the price of these models is high. Cheaper models are available in the U.K., but these are not as accurate.

(4) *Electrocardiogram.* The vet can examine the heart using an electrocardiogram (ECG), which is a recording of the electrical activity caused by nerve impulses in the heart muscle. This is a very specialised technique which has been used to relate the horse's heart size to performance potential. Assuming that a larger mass of heart muscle requires a longer time to contract, a specific portion

of the ECG trace is thought to indicate the heart size of the animal, and is called the 'heart score'. The theory is that the larger the heart in relation to the horse's size, the better able it will be to meet the oxygen demands of the muscles during exercise. A study in Australia found that racehorses with the best performances had higher than average heart scores, i.e. bigger hearts. It is suggested that heart score can be used to indicate superior athletic ability.

A slower resting heart rate also indicates a larger heart. The bigger heart has to do less work to pump the same amount of blood, thus a fit horse with a well-developed muscular heart has a slower resting heart rate than an unfit horse.

It must be remembered that heart rate can vary considerably between individual horses, and each rider should make a conscientious effort to know the resting heart rate for his own horse. Once this is established, the heart rate can be used to monitor the horse's progress during a conditioning programme.

The Heartbeat

The actions of the heart during one complete heartbeat make up the 'cardiac cycle'. Simultaneously blood enters the right atrium from the body and the left atrium from the lungs. The atria contract when they are full, opening the valves and thus allowing blood to flow into the relaxed ventricles. When nearly full the ventricles begin to contract, closing the valves to the atria. While the ventricles contract blood is forced out of the heart, through non-return valves, to the body and lungs. These are the diastole and systole actions of the heart, 'diastole' being the relaxation (dilation) of a chamber just before and during its filling and 'systole' the contraction of a chamber in the process of emptying.

When the heart beats it makes sounds. There are actually four heart sounds, but only two are loud enough for the average ear to distinguish. Commonly known as 'LUBB-dup', they mark the opening and closing of valves during the sequence of heart action. The 'LUBB' sound marks the point in the beat when the valves between the atria and the ventricles are closing and the artery valve exits are opening to permit the flow of blood to the lungs and body. The crisp-sounding 'dup' occurs as the inter-chamber valves open and the arterial valves close.

The heart beats because of electrical impulses that come from a natural built-in pacemaker. Signals for controlling heart rate come from the brain via nerves. At regular intervals the pacemaker sends out an electrical impulse which causes a wave of contraction in the heart muscle, and the heart goes through an orderly series of systoles and diastoles as described above. The silent period in the heartbeat occurs when the heart is filling with blood. The sounds the heart makes can indicate the health of the heart.

Heart Murmurs and Arrythmia

Unlike human heart attacks, most sudden equine deaths result from rupture of the aorta, the main artery which carries oxygenated blood from the heart. The human heart-attack victim collapses because the muscle of the heart wall has had its blood supply cut off; the coronary artery has been blocked, e.g. by a clot, and this causes part of the heart muscle to die. The heart is unable to beat, resulting in collapse.

Although horses rarely die through cardiac problems, normal resting horses' hearts often skip beats and fit, functional horses may develop loud variations of normal heart sounds. Fortunately, however, the majority of horses whose hearts make odd sounds or beat erratically function quite well. It has been suggested that over half of all normal healthy horses may have innocent murmurs and benign arrhythmias.

Murmurs

A murmur is an unusually loud or odd sound that can be heard during what would normally be a silent period in the cardiac cycle. Murmurs are the most common type of abnormality found during routine examination, and may be described as whistling, harsh, blowing, machinery-like, buzzing, squeaking, high-pitched, cooing or hissing – they all indicate turbulence in blood flow through the heart. Some are caused by changes in the pumping action or in the consistency of the blood, but these are not generally serious. Serious murmurs result from structural flaws in the heart.

Fit athletic hearts are prone to murmur because training builds up heart muscle so that the heart pumps more blood with each stroke. Thus the resting heart rate is slower and the resulting rush of blood is louder than normal. Technically this sound is a murmur, but functionally it is a normal noise for the heart.

Other murmurs that occur in horses with normal hearts can be explained by changes in blood circulation brought on by fever, stress and anaemia. Increased body temperature and stress speeds up heart rate and blood flow, which can lead to abnormal sounds in the cardiac cycle. Anaemia encourages odd heart sounds because the number of red blood cells is reduced, altering the consistency of the blood. The turbulence of

normal

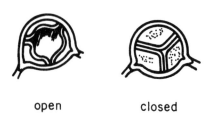

open   closed

faulty

open (insufficient)   closed (incompetent)

Fig. 6.10 Normal and faulty heart valves

the thinner fluid passing through the heart is heard as a murmur. Most such murmurs produced by normal hearts will subside as the horse regains health.

Murmurs can arise when the valves in the heart, which separate the chambers, deteriorate with wear and tear (fig. 6.10). The valves become shortened and thickened so that they close imperfectly and blood can leak through. This leaking blood makes a noise heard through the stethoscope as a murmur. Older horses often manage quite well with 'loud valves', but it is rare for these older horses to be pushed for maximum performance, which is when valve problems could show up.

Listening to a horse's heart sounds both at rest and after exercise can indicate the severity of a murmur. The two main points the vet will listen for are:

(1) The timing of the murmur in the cardiac cycle.
(2) How the murmur reacts to exercise. Murmurs that disappear or fade with exercise are not usually serious.

Low pitched 'blowing' murmurs that occur with contraction (following the 'LUBB' sound) are usually the harmless sounds of the rush of blood ejected from a larger than average heart. Loud heart murmurs that hide the normal crisp valve sounds (the 'dup') usually indicate a valve malfunction. Flow sounds during the heart's rest period indicate that blood is leaking back from the exit arteries.

Arrythmias

Occasionally the resting horse's heart will skip a beat. In a fit horse, this is probably a way of regulating total blood supply to the tissues. Even though the beat is as slow as it can go, the heart is still ejecting more blood per minute than the horse's needs at rest. The solution is to miss a beat occasionally. When the horse is exercised and the demand for blood increases, the heart will beat regularly.

'Random arrythmia' is an irregular flurry of rapid beats. When this occurs the heart is not functioning efficiently: a slow steady stroke allows refill time, but rapid irregular beats do not. This flutter occurs due to a failure in the nervous control of the heartbeat. The nerve centre in the heart which stimulates contraction (the pacemaker) may have had to move because the original site has been damaged in some way, e.g. as a result of inflammation. Additional pacemakers may have arisen which compete with the original pacemaker, and thus random signals could be causing the heart to contract before the chambers are filled with blood.

The most common arrythmia is called 'atrial fibrillation', and occurs in the upper chambers of the heart. During exercise, the fibrillating heart may beat up to 260 times per minute, as compared to a normal maximum of 240 times per minute. A sudden drop in the horse's performance is one sign of the onset of this condition. Treatment with drugs combined with rest may return the heart beat to normal. Heart muscle inflammation brought on by infective damage may cause changes in the pacemaker, and may follow an attack of equine influenza.

Although it would appear that a number of horses die with heart abnormalities, very few die *because* of them. Murmurs and arrhythmias are often conditions that a horse can live with, provided it is not expected to perform at very high levels. Generally speaking, the horse's heart copes with the demands put on it very well.

The Blood

Blood is the vital transport medium connecting the heart and the lungs to the rest of the body. It carries nutrients absorbed from the digestive tract to all tissues of the body. Oxygen is transported from the lungs to the tissues, and carbon dioxide is carried back from the tissues to the lungs. The blood carries waste products from the cells to the kidneys for excretion. Hormones and factors by which the body fights disease are also carried in the blood. The clotting ability of blood helps insure that large

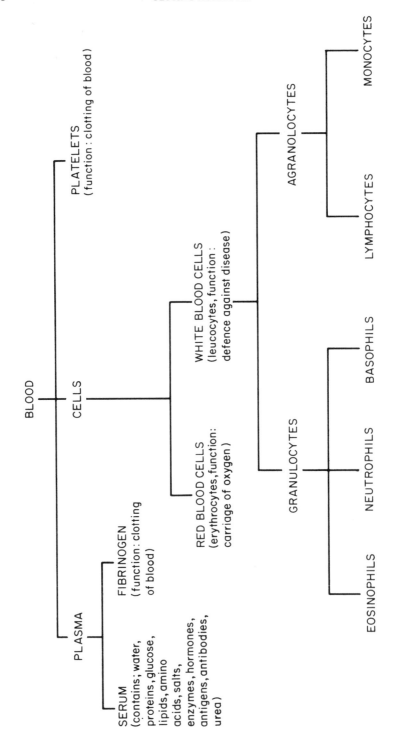

Fig. 6.11 The components of blood

quantities of blood are not lost through injury. In addition, blood helps control body temperature by transporting heat from deep structures to the surface of the body.

The Structure of the Blood

Plasma

The composition of the blood is illustrated in fig. 6.11. It consists of a straw-coloured fluid called *plasma* in which all the cellular components of the blood are suspended. These are red cells, white cells and platelets. Plasma contains *fibrinogen* which aids clotting of blood, and *serum* which contains water, proteins, glucose, lipids (fats), amino acids, salts, enzymes, hormones, antigens, antibodies and urea. The proportion of cells to plasma, known as the 'packed cell volume' (PVC), is the range of 31 to 55%, 40% being an average value.

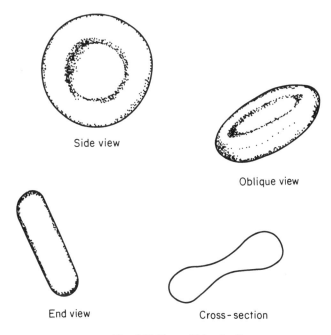

Side view

Oblique view

End view

Cross-section

Fig. 6.12 The red blood cell

Red Blood Cells

Red blood cells or *erythrocytes* originate in the bone marrow. They are tiny cells containing haemoglobin, the respiratory pigment. Their job is to carry oxygen; when exposed to high concentrations of oxygen, e.g. in the lungs, haemoglobin takes up oxygen and carries it as oxyhaemoglobin. In the tissues there is very little oxygen, and here the haemoglobin releases the oxygen so that the tissues can use it to produce energy.

Red blood cells are structured with thick edges and thin centres, providing a larger surface area for oxygen exchange (fig. 6.12).

These cells are continually being worn out and replaced. Worn out cells are destroyed in the liver, and products which cannot be used again are excreted in the bile.

A Thoroughbred-type horse normally has 9 to 12 million red blood cells per cubic millimetre of blood.

White Blood Cells

There are five kinds of white blood cell or *leucocyte:* neutrophils and basophils, eosinophils, monocytes and lymphocytes (fig. 6.13), their main function being to defend the body against disease. Some, such as

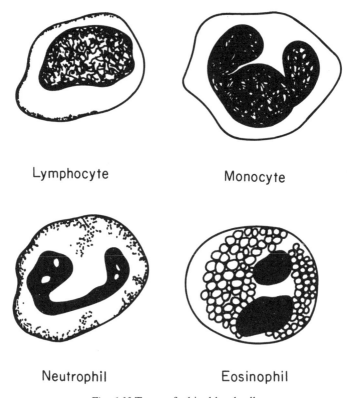

Lymphocyte   Monocyte

Neutrophil   Eosinophil

Fig. 6.13 Types of white blood cell

neutrophils, are most active in defence of acute infection while others, such as monocytes, are most active in the face of less acute infections.

Leucocytes are carried by the blood to general locations of infection and then can move to the actual site of the infection. Leucocytes fight infection by engulfing the invading bacteria and digesting them. This material, with dead tissue, is what constitutes pus.

Blood Proportions

Each type of cell is present in the blood in characteristic numbers. These numbers can be used to assess the health and fitness of a horse when a blood test is taken, as will be discussed in more detail later.

The volume of blood present in the horse is about 10% of its body weight, i.e. a typical 500 kg horse has about 45 litres (10 gallons) of blood. About 60% of this is water and the blood plays a vital role in regulating the fluid balance in the horse's body. This can be of importance in a performance animal that sweats heavily. Not all the 45 litres of blood is circulating round the horse's body all the time. Some blood is stored in the spleen, an organ situated near the stomach which acts as a reservoir of blood. Blood is released from the spleen in times of stress, e.g. during exercise. The release of these splenic reserves can alter the values obtained in a blood test dramatically.

The Effects of Training on the Blood

Red Blood Cell Number

When considering the horse as an athlete, great importance is placed on the number of red blood cells because of their oxygen-carrying capacity. Without an adequate oxygen supply to meet the requirements of working muscle, the equine athlete cannot meet the demands of competition. The packed cell volume (P.C.V.) and red blood cell count (R.B.C.) indicate the oxygen-carrying capacity of the blood and have been found to increase with training. However, the effect of the release of splenic reserves of red blood cells on R.B.C. and P.C.V. must be remembered. Age, sex and breed will also cause variation.

Rate of Turnover of Red Blood Cells

Another important effect of training on the blood is an increase in the rate of turnover of red cells. Cells have a limited lifespan, usually four to five months. However, as a horse works harder the lifespan of the red cells decreases. Cells travel hundreds of kilometres through the blood vessels and eventually a combination of wear against the walls of the blood vessels and squeezing to pass through the tiny capillaries literally wears the cells out. A working horse pumps the blood through the body more quickly, and so the process of wearing out takes place more quickly too. Younger red cells have a greater ability to carry oxygen, so a fit horse with a younger average population of red cells has a greater oxygen carrying capacity.

Blood Testing

The vet can collect blood samples from a horse, have it tested and have the results ready in a couple of days. This makes blood-testing an invaluable tool in helping the vet confirm a diagnosis of illness or assess well-being and fitness to compete. For a blood test to be of any use in determining deviations from the normal, i.e. illness, then the normal for that horse must be known. Blood tests can only be of maximum use if taken *regularly* as part of the horse's training programme.

Blood Collection

The skin, usually over the jugular vein, is swabbed with surgical spirit and

Fig. 6.14 Blood sample collection

a needle is inserted into the vein (fig. 6.14); 2 to 3 ml of blood is removed and mixed immediately with an anticoagulant to prevent clotting. A further 5 to 7 ml is taken and placed in a separate tube and allowed to clot for serum examination.

Few horses appear to resent this procedure – much less so than an injection of antibiotic. It is important that the horse is upset as little as possible and that it has not been exercised that day as some values change after stress and exercise. The sample should reach the laboratory within 24 hours of collection because blood deteriorates if kept too long.

Blood Analysis

The unclotted blood is placed in special microscope slides with counting chambers and the total red cells are counted by sophisticated electronic equipment. The haemoglobin content of the blood is estimated. Unclotted blood is then spun at high speed (centrifuged) so that the cells all pack in one end of a parallel-sided tube. By measuring the length of cells in the tube, the P.C.V. percentage is calculated. This test indicates that a horse is dehydrated if there is a larger than normal P.C.V. Along with red cell count and haemoglobin concentration, P.C.V. can be used to calculate the mean corpuscular haemoglobin concentration (M.C.H.C.). These tests can indicate anaemia due to deficiences of iron, folate or vitamin B_{12}.

A white cell count is also carried out. Any deviation from the normal numbers of white cells is usually because a greater demand is being made upon them, e.g. increased neutrophil numbers indicate bacterial infection and/or inflammation. Lymphocytes produce antibodies to specific diseases and germs; those that react to viruses are different and can be seen in a blood film, often before the illness is detectable. Eosinophils multiply in allergies (an intolerance to foreign protein); eosinophil numbers also increase if the horse is suffering from a worm burden. Monocytes can increase as part of an immune reaction, or when bacterial infection is present, or if there is an internal parasite infection.

White cells are counted as a whole and the different types are also counted.

Serum Examination

When blood is allowed to clot, the remaining fluid is called serum. When anti-clotting agents are added and the cells removed, the remaining fluid is called plasma. Plasma consists of serum plus the agents involved in clotting. Serum is examined for proteins and enzymes (biological catalysts that enable metabolic processes to take place), a process called serum biochemistry.

Serum proteins are of two main types:

• albumin
• many globulins

Albumin is made in the liver when the diet is adequate, and globulins are produced by the immune system and are known as 'antibodies' to disease. As the animal ages, more diseases are encountered; antibodies increase and so globulins and the total blood protein increases. Younger animals naturally have lower total protein in the blood, and this cannot be increased by feeding more protein.

Low albumin usually indicates that the diet is poor or that albumin is being lost, as during worm infestation.

Correct functioning of organs of the body can be tested. For example, liver function is checked by protein estimations and serum enzyme tests; kidney function is tested by measuring serum protein and serum urea. Muscle is rich in an enzyme called creatine phosphokinase (C.P.K.), and damage to muscle causes an increase in C.P.K. levels. Testing for this enzyme helps in diagnosis and treatment of azoturia.

With specialised equipment, the blood can also be tested for a full range of minerals. It may be wise to have a horse's blood tested for selenium if you live in a selenium-deficient area.

It must be remembered that each horse is different – what is normal for one horse may be unusual for another. Ideally, not only should the normal for each horse be known but the results of the test should be compared to a table of the laboratory's normal ranges of results (table 6.1).

Interpretation of the Blood Test

This is the job of the veterinary surgeon (see Fig. 6.15), requiring experience and professional expertise. However, some signs can be looked for:

(1) *Infection*
    (a) Bacterial
        • increased total white cell count
        • an increased percentage of neutrophils
    (b) Viral
        • increased total white cell count
        • more lymphocytes
        • presence of virus lymphocytes

Table 6.1 Normal range of blood test values

Haematology

Red cell count	8.5–11 million/mm^3
Haemoglobin concentration	13–17 gm/100 ml
Packed cell volume	34–44%
Mean corpuscular volume	38–45 Femto litre
Mean corpuscular haemoglobin concentration	32–39 gm/100 ml
White cells	6 000–12 000/mm^3
Neutrophils	2 000–8 000/mm^3
Lymphocytes	1 500–4 000/mm^3
Eosinophils	100–600/mm^3
Monocytes	100–600/mm^3
Basophils	20–50/mm^3

Biochemistry

Serum proteins	
total	55–75 gm/l
albumin	25–41 gm/l
globulins	25–41 gm/l 1:1 ratio
Urea	20–45 mg/100 ml
C.P.K.	20–80 iu/l

iu = international unit

---

(2) *Anaemia*
- reduced red cell count
- low haemoglobin concentration

Anaemia is usually found in the following circumstances:
- poor nutrition
- worm burden
- deficiences of iron, vitamin B$_{12}$, folic acid

(3) *Dehydration*

This is a lack of body fluid caused by fluid loss being greater than fluid intake (e.g. after illness, exertion, water deprivation), and is shown by the P.C.V.

(4) *Muscle damage*

May result from injury, exertion (especially during dehydration or illness) or azoturia, and is shown by high C.P.K. levels.

```
LABORATORY BLOOD TEST REPORT

OWNER: NAME...............................

       ADDRESS............................

       ...............................

       ...............................

VETERINARY SURGEON:......................

Date Received...........................
```

	NAME OF HORSE:	RAGROW
	LAB. REF. NO:	
HAEMATOLOGY		
Red Cell Count		10.80
Haemoglobin g/100ml		14. 5
PCV %		43
McHc %		34
White Cell Count		7,900
Polymorphs %		55
Lymphocytes %		42
Eosinophils %		2
Basophils %		
Monocytes %		1
Erythrocyte Mins 10		
Sedimentation 20		
30		
60		
BIOCHEMISTRY		
Serum Protein: Total g/100ml		6.8
Globulin g/100ml		4.0
Albumen g/100ml		2.8
SGOT		
CPK		
Serum Calcium mg/dl		11.3
Phosphorus mg/dl		3.0
(Ca:PO4)		(3.8:1)

Fig. 6.15 Blood test results

7   Bone and Hoof Structure

Bone

Structure and Function

Bone is the hard substance making up the body's skeleton, on which the soft tissues hang. Bone is composed partly of living fibrous tissue and partly of non-living calcium and phosphorus salts. The living tissue is called collagen, a fibrous protein also found in tendons and cartilage. The collagen is organised into a matrix encrusted with mineral crystals, mostly calcium phosphate. The proportion of living to non-living (mineral) material in bone varies according to the age of the horse. Young horses have soft bones consisting of 60% fibrous tissue, whereas adult animals have more brittle bones containing only 35% fibrous tissue.

Although bone appears to be hard, dense, inelastic and almost lifeless, its structure is constantly changing – the entire calcium content of the skeleton is replaced every 200 days. Bone serves as a mineral reservoir and is being constantly replenished or depleted. No tissue in the body is capable of as much overgrowth and as much absorption as bone.

Bone is of two types: dense and spongy. Dense bone is found in the shafts of the long bones of the limbs. Under the microscope, dense bone has a tube-like form with a central cavity containing fatty yellow marrow. The outer surface is covered by the bone membrane, or periosteum. Spongy bone is found in the short bones and at the ends of long bones. It has a more open framework than dense bone and contains red bone marrow. Red blood cell synthesis takes place in red bone marrow.

All bone is penetrated by a series of very fine canals in which blood vessels, nerves and lymph vessels lie. These bring the stimuli and nutrition necessary for growth, maintenance and repair of bone.

Bone has many functions in the body other than providing support and rigidity. Bones are the levers that give muscles something to work against; they protect internal organs and act as stores of calcium and other minerals. The mechanical function (support, leverage or protection) determines the shape of each bone, leading to bones being classified by shape.

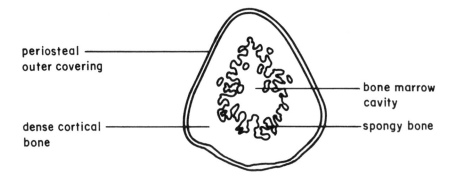

periosteal
outer covering

bone marrow
cavity

dense cortical
bone

spongy bone

Fig. 7.1 A cross-section through a long bone

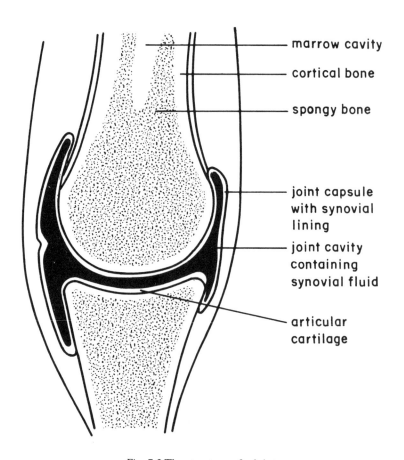

marrow cavity

cortical bone

spongy bone

joint capsule
with synovial
lining

joint cavity
containing
synovial fluid

articular
cartilage

Fig. 7.2 The structure of a joint

(1) *Long bones.* These are main supporting columns and levers, e.g. the cannon bone, being cylindrical shafts with enlarged ends called epiphyses.

(2) *Flat bones.* These act as protectors, e.g. the skull which protects the brain.

(3) *Short bones.* These act as shock absorbers, e.g. the small bones of the knee and hock joints.

(4) *Irregular bones.* These are bones with a special role to play, e.g. the vertebrae.

Regardless of their shape, the composition of these bones is similar. A cross-section of a long bone (fig. 7.1) shows an outer layer of dense material called 'compact bone' which is thickest where stress is the greatest, nearest to the outer surface of the cylinder. There is an inner layer of spongy bone which is interspersed with marrow. Marrow is involved with the production of blood cells. Long bones have spongy bone at either end while the interior of the shaft is hollow and filled with marrow.

Bones are surrounded by the periosteum, which acts as an attachment point for the ligaments and tendons. The periosteum is lined with cells called osteoblasts, which produce new bone. The periosteum also brings nutrients to the bone via blood vessels and nerves that enter the bone through a series of canals.

The joint surfaces of the bones have a cap of shiny resilient cartilage (fig 7.2), which provides a smooth articulating surface and lubrication in the form of synovial fluid.

Formation and Growth

In the unborn foal, bone begins as cartilage. Bone-forming cells (osteoblasts) invade the cartilage and begin to secrete the matrix; minerals are deposited on the matrix and it calcifies into bone. By the time the foal is born, calcification is complete in most bones (fig. 7.3).

Growth of Long Bones

Bones have to increase in both length and diameter. Increase in length takes place at two narrow bands of cartilage, one near each end of the bone. These bands are called the epiphyseal 'growth' plates (fig. 7.4). Cartilage grows on the side of the growth plate nearest the end of the bone, while the shaft side of the growth plate is invaded by osteoblasts which turn the cartilage into bone. Gradually the osteoblasts gain on the cartilage, and when the bone has reached its proper size the plate closes

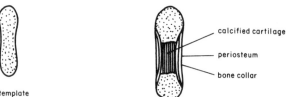

1. The early cartilagenous template

2. Formation of periosteal bone collar

calcified cartilage

periosteum

bone collar

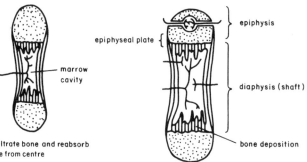

marrow cavity

3. Blood vessels infiltrate bone and reabsorb calcified cartilage from centre

epiphyseal plate {

epiphysis

diaphysis (shaft)

bone deposition

4. Blood vessels invade epiphysis, resorbing cartilage and forming a growth centre at the bone's end. Osteoblasts deposit bone in the epiphyseal plate

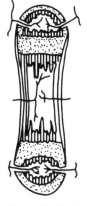

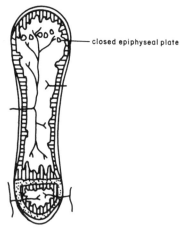

closed epiphyseal plate

5. In epiphysis cartilage replaced by bone. At other end same transformation. Shaft lengthened by bone deposited at epiphyseal plate

6. At bone's upper end all cartilage (except joint cartilage) replaced by bone. Bone has grown upward across epiphyseal plate and the marrow of the epiphysis is continuous with that of the shaft. Growth at that end of bone is ceased

Fig. 7.3 The growth and development of a long bone

and becomes bone. Different bones stop growing at different times; for example, growth plates of the cannon bone close at nine to twelve months and those of the tibia at three and a half years.

Bone increases in diameter through the action of the osteoblasts lining the periosteum. These cells lay down new material in layers over the old bone, giving a ring-like effect similar to the rings of a tree trunk. Simultaneously, old bone is destroyed by osteoclasts so that the marrow filled cavity grows and the shaft wall is only slightly thicker. Osteoclasts break up dead and damaged bone and reduce the size of any callus or thickening where a mend occurs.

Adaptation of Bone to Stress

After growth stops, bone undergoes a process called 'remodelling'. This involves a balance between the formation of new bone and the reabsorption of old bone, and has two main functions: it allows the use of mineral stores and enables adaptation to stress.

Remodelling starts when the foal is about three months old, when the newly formed bone begins to rearrange itself into 'Haversian systems'. These systems consist of a series of vertically aligned tubules through

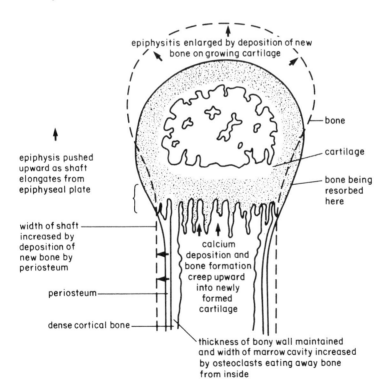

Activity at the epiphyseal plate not only adds length to the shaft, it also enlarges the epiphysis. Developments at the periosteum increase the diameter of the shaft.

Fig. 7.4 The activity of the epiphyseal plate

which the blood vessels travel. Remodelling is usually complete by the time the horse is six years old. If the Haversian systems are formed too quickly with too little mineral content, the bone becomes porous, and porous bone is not as strong as dense bone. This will happen if the calcium supply in the diet is not adequate, so a correct balanced diet is essential for the development and maintenance of a healthy skeleton.

The Effect of Exercise on Bone

Although bone appears lifeless, it is extremely responsive to environmental changes. Changes in pressure, blood supply and nutrition can all bring about changes in bone structure. Bone can decrease in size (atrophy), increase in size (hypertrophy), repair breaks and rearrange its internal structure to resist stress and strain. Bone is able to reshape itself to sustain a maximum of stress with a minimum of bone tissue.

Atrophy of bone occurs where constant and excessive pressure is applied or where periods of pressure exceed periods of release. Proliferation or hypertrophy of bone occurs in response to concussion or intermittant pressure. Whether bone atrophies or proliferates under pressure is largely dependent on the degree and duration of the pressure and on the maturity of the bone. Pressure on growing bone will slow down and may even stop growth. Mature bone responds to pressure by excess growth or rearrangement of its structure. These changes often exhibit themselves as problems such as splints, ringbone and other bone diseases, and will be discussed later.

Short bones such as the bones of the knee joint are not as resistant to compression and pressure as are long bones such as the cannon bone. Thus short bones tend to develop bony growths or hairline fractures which are painful, restrict movement and often will not heal. The bones of the foot are short bones. The navicular bone is sandwiched between the pedal bone and the short pastern bone. Both these bones are very much larger and very much more dense than the navicular bone which is subject to damage from crushing forces sustained during galloping and jumping, contributing towards navicular disease.

During exercise bone is subjected to stretching forces (tension). Bone is relatively inelastic – a rod of bone can be stretched by about 0.005% of its length before breaking. If stretched to near breaking point, bone will not return completely to its original length. Under stress bone deforms and does not return to its original length – an exaggerated example of this is seen in rickets.

Bone is also subject to compression, shearing, bending and twisting during movement. A bone will support more weight when the animal is stationary than it can bear under a dynamic or moving load. The leg bones of a horse bear a static or stationary load when the horse is standing still and a dynamic load when the horse is galloping and jumping. Movement causes compression, bending and shearing stresses. When a horse pivots with one foot in contact with the ground, twisting stresses are added to the bone. Muscles and tendon act like 'guy wires' to reduce stresses of all types on the bone. To do this effectively, the muscles and tendons must themselves be capable of withstanding the stresses: they must be fit.

Bone responds to the stress created by exercise by remodelling and laying down more bone so that the mass or weight of the skeleton is greater. Conversely, if bone is not subjected to stress it is reabsorbed and the mass of the skeleton decreases. Every time a horse is ridden the bone is stressed and the skeleton undergoes changes. This means that when we condition a horse's muscles we are also conditioning the skeleton. The skeleton is the foundation of the performance horse and it is important to be aware of the effect that work is going to have on the horse's bones.

Bone is a rigid structure and cannot be moulded. The only way that it can change shape is by adding more bone to its surface or by taking bone away. Bone can only repair damage or change its internal structure by the osteoblasts 'drilling' a hole towards the damaged area and the osteoclasts filling in the hole behind them. It would seem that during repairing bone actually goes through a weak stage, because the first stage of repair is to *remove* bone, thus weakening it, before the bone-making cells can go in and repair the damage. If bone is damaged during training and training continues, a vicious circle is set up: less bone, more damage, more reabsorption, less bone, etc. This can eventually lead to stress fractures (fig. 7.5) – the bone breaks due to repetition of stress in much the same way that a piece of wire breaks after being bent repeatedly. In a horse these stress fractures can be seen as 'sore shins', a consequence of training young horses on hard ground.

Controlled amounts of stress will benefit the bone by stimulating the remodelling process, resulting in more dense, stronger bone. A training programme should contain periods of brisk trotting and exercise should be as diverse as possible, certainly including all types of stress that the bone will be subject to during competition. It is beneficial for exercise to be vigorous, thus inducing high levels of strain, but these levels should not continue for long periods.

It takes about three months of work before remodelling has significant effects, but the process goes on for many months and the effect will vary between individuals. One study of professional tennis players found that on the serving side of their body they had an average of about 35% more

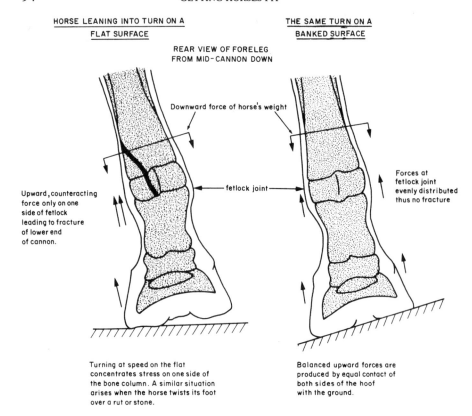

HORSE LEANING INTO TURN ON A
FLAT SURFACE

THE SAME TURN ON A
BANKED SURFACE

REAR VIEW OF FORELEG
FROM MID-CANNON DOWN

Downward force of horse's weight

Upward, counteracting
force only on one
side of fetlock
leading to fracture
of lower end
of cannon.

fetlock joint

Forces at
fetlock joint
evenly distributed
thus no fracture

Turning at speed on the flat
concentrates stress on one side of
the bone column. A similar situation
arises when the horse twists its foot
over a rut or stone.

Balanced upward forces are
produced by equal contact of
both sides of the hoof
with the ground.

Fig. 7.5 How a stress fracture can occur

bone than on the non-serving side, some having as much as 100% more. The stress a horse's bones receive has been divided into three types:

(1) *The elastic zone.* The normal elastic properties of the bone allow it to absorb the impact of 900 to 1350 kg (2000 to 3000 lb) each time the forelegs hit the ground (an easy canter) with no change to its structure.

(2) *The fatigue zone.* The bone does not return to its original state after each stride, and slow deformation occurs. The fatigue zone is reached during fast galloping when each foreleg experiences about 4000 kg (9000 lb) force as it lands. When the horse rests the bone repairs itself and is stronger. However, if there is insufficient time for healing before the next gallop, the bone may become permanently damaged – up to four weeks may be necessary. If the skeletal system is to be made significantly stronger, the horse must undergo work in the fatigue zone.

(3) *The overload zone.* The bone has no elasticity left in it; it undergoes

severe deformation, and snaps. This occurs at about 5500 to 8000 kg (12,000 to 18,000 lb) force, which can happen if the bone is loaded unevenly – for example, round a turn where the entire force may fall on one side of the cannon bone (see fig. 7.5).

The changes in density of bone can be measured by the velocity of ultrasound waves sent through the bone, a high velocity indicating strong, dense bone. This technique has been used extensively in the U.S.A. to demonstrate the effect of exercise on bone.

Although it is not definitely known what it is in the stress that signals to the bone cells to produce more bone, it is thought to be electrical in origin. Mild electric currents induced by magnetic fields have been shown to speed the healing of broken bones in humans. This idea has been introduced into equine therapy to aid healing of bone-related conditions.

It can be seen, then, that bone can be conditioned just like the other body systems, and it must be stressed in order to sustain the loads put upon it by the demands of performance.

Diseases of Bone

Bones are deeply seated within other tissues and are not richly supplied with blood, which means that bone disease is often overlooked.

Diseases of bones fall into five main categories:

- periostitis
- osteitis
- disorders of the growth plates (epiphysitis)
- infection
- fracture

Periostitis is inflammation of the surface of the bone and its covering surface (periosteum). It is most usually caused by sprain, blow or infection and the symptoms are pain, swelling and heat over the affected area. Examples of conditions caused by periostitiss are ringbones, osselets, sore shins and spavins.

Ringbones involve the growth of new bone (exostosis) on the pedal bone, short pastern or long pastern bone. The condition is caused by trauma (e.g. a blow or tread), underlying bone disease, nutritional deficiency or infection. Ringbones are classed as high or low depending on where they are sited.

Osselets involve growth of new bone on the front of the fetlock. The condition is caused by concussion leading to inflammation of the periosteum of the fetlock joint.

Sore shins arise due to the lining of the cannon bone being inflamed, caused by galloping on hard ground. The periosteum becomes detached and new bone is formed.

Spavin is a condition of the hock joint. Bone spavins are caused by periosteitis or osteitis of the upper end of the cannon bone and some of the small bones of the hock, causing lameness.

Osteitis involves inflammation of the bone substance itself – the most common complaint is pedal osteitis, where the pedal bone is inflamed causing pain and lameness.

Epiphysitis occurs when the growth plate becomes inflamed. This may be caused by infection, trauma or an imbalanced diet (e.g. excess protein or an incorrect calcium/phosphorus ratio). It is most common during periods of rapid growth – for example, in the lower end of the radius (the bone immediately below the knee) at 12 to 24 months. The symptoms are pain, swelling and lameness.

Infection of bone can be very serious, and may require surgical intervention.

Fractures of bones can be classified according to the age of the animal and the type and site of the break. Fractures of small bones (e.g. splint) may not be serious, while breaking a large weight-bearing bone can result in the horse having to be destroyed.

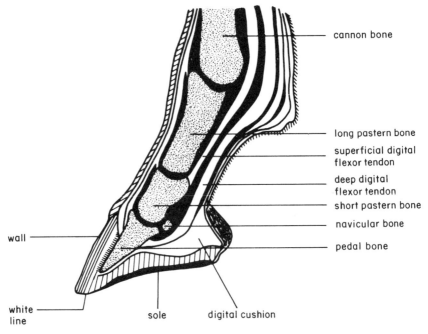

Fig. 7.6 The structure of the foot, showing bones and tendons

Splints may arise from fractures of the splint bone resulting in a callus. The cause may be a knock, as when one foreleg strikes another.

Many of the diseases described here arise as a result of concussion or trauma. In other words, they are exercise induced injuries and as such they are common in performance horses. Bone is well able to adapt and thus withstand the stress of the horse moving; it is only when the stress becomes excessive or prolonged that these diseases arise. By understanding the processes that cause the stress and the response of the bone to that stress, the rider should be able to avoid placing his horse in a position where the stress becomes excessive. Developing the horse's muscle, bone and tendon by a planned, systematic training programme will help the horse adapt to the demands made on its body and help minimise the risk of these injuries and diseases occuring.

The Hoof

The foot of the horse (fig. 7.6) consists of a horny box which surrounds the pedal bone, the navicular bone, ligaments, tendons, the digital cushion, sensitive laminae, blood vessels and nerves. The horny box is called the hoof and consists of a wall divided into toe (at the front) and quarters (at the side). The sole, bars and frog are underneath. The wall is bent sharply backwards at the heels to form the bars.

The hoof wall is manufactured by specialised cells of the skin (corium)

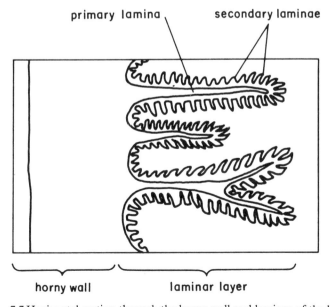

Fig. 7.7 Horizontal section through the horny wall and laminae of the hoof

and, in effect, is modified skin or nail. The wall is about 25% water and is protected from evaporation and thus drying by an outer layer called the periople. The middle layer of the wall is dense and hard and contains pigment in dark-coloured feet. The inner layer of the wall is called the laminar layer and forms specialised membranes of laminae which attach the hoof to the pedal bone (fig. 7.7). There are hundreds of tiny primary laminae, each of which has about 100 secondary laminae. These interdigitate or 'dovetail' with sensitive laminae which cover the outside of the pedal bone. This means that, although each small pedal bone supports about a quarter of the horse's weight, the burden is spread over a very large surface area. If this were not the case, the pedal bone would be pushed through the sole of the foot as in laminitis.

On the ground surface, the wall is united to the sole by horn of a lighter-colour and softer texture, called the 'white line'. This marks the boundary of the sensitive laminae, and thus shoeing nails must be outside this line. The sole is the ground surface of the hoof, and is concave in shape and bounded by the walls and bars. The frog is a wedge of soft horn between the bars and is the most elastic structure of the foot. Inside the hoof, above and around the frog, covering the heel, is some soft tissue called the digital cushion or plantar cushion.

On either side of the pedal bone are two plates of cartilage called the lateral cartilage (fig. 7.8). The frog, digital cushion and lateral cartilages act together as an anti-concussion mechanism and an aid to blood circulation in the foot and leg. As the foot strikes the ground and the horse's weight travels over it, the frog is forced into the bars, the heel expands and the forces of concussion are distributed. The digital cushion is compressed against the frog and lateral cartilages and the lateral cartilages move outwards, forcing blood out of the veins and capillaries trapped between the cartilages and the wall. This has two effects: firstly, the blood is forced out of the foot up the leg, and secondly, the blood confined within the hoof wall has a hydraulic shock-absorbing effect.

If the horse is confined to its box, this forced return system of blood is inactive, resulting in a poor blood flow and filled legs. Nails placed too far back in the quarters of the foot inhibit the elasticity of the heels and their ability to expand, and thus affect blood flow within the foot. Frog contact is also very important in ensuring shock-absorption and correct blood flow in the foot.

Growth of the hoof takes place at the coronet, at a rate of approximately 0.6 cm (0.25 in) a month, taking nine months to grow from coronary band to toe. Growth of healthy hoof reflects the state of health and the plane of nutrition of the horse.

Heart and lung development are vital to the well-being of the feet,

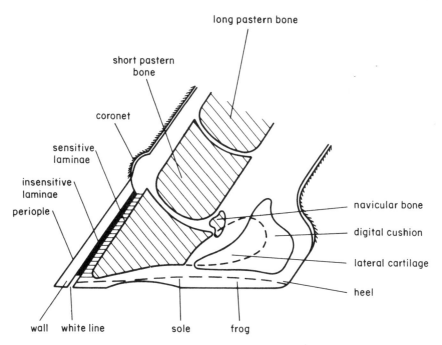

long pastern bone

short pastern
bone

coronet

sensitive
laminae

insensitive
laminae

periople

navicular bone

digital cushion

lateral cartilage

heel

wall   white line   sole   frog

Fig. 7.8 The structure of the foot, showing the digital cushion and lateral cartilage

because the structures of the foot require a large volume of blood. This blood is needed for nourishment of tissue, maintenance of warmth and dissipation of heat. Many diseases of the foot may be partly attributable to poor circulation – for example, navicular and sidebone.

Exercise will have an effect on hoof growth. As the horse works, its heart rate increases and blood is pumped more rapidly round the body. The blood supply to the foot and coronet is increased, bringing with it a supply of nutrients for growth; thus hoof growth increases. Simultaneously, the plane of nutrition of a working horse improves as feed is increased, supplying the nutrients necessary for increased hoof growth. By looking at the growth rings on the hooves of a hunter it is possible to show quite clearly when the horse was brought into work and when turned away.

8 The Dental System

Teeth are the beginning of the digestive system of the horse and they have a very important role to play. The horse has two sets of teeth during its life:

(1) *Temporary* or milk teeth, smaller and whiter than permanent teeth.
(2) *Permanent* teeth, larger and yellower adult teeth.

The horse has only three types of teeth:

- incisors or biting teeth in the front of the mouth (fig. 8.1)

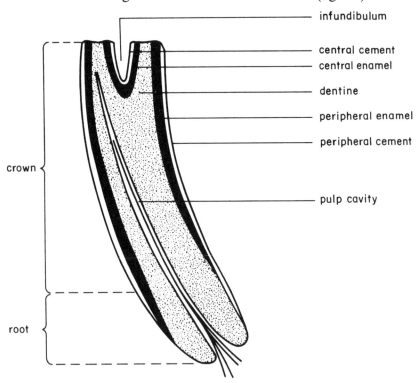

Fig. 8.1 The incisor tooth

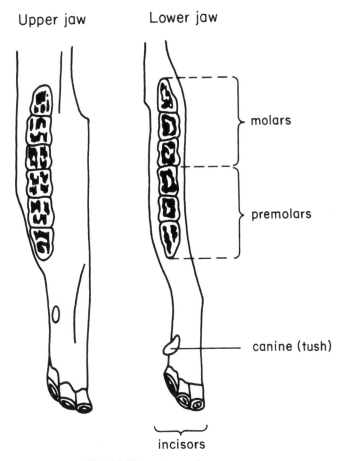

Upper jaw   Lower jaw

molars

premolars

canine (tush)

incisors

Fig. 8.2 The molar teeth

- molars or grinding teeth lining either side of the jaw bone – the cheek teeth (fig. 8.2)
- tushes or canine teeth

There are six incisors in each jaw: two corners, two laterals and two central incisors. These teeth are used to age horses. There are twelve molars to each jaw, six each side. A horse has an adult mouth of permanent teeth as a five-year-old. Adult male horses have two tushes in each jaw, situated between the incisors and the molars, but mares rarely have tushes – thus mares have a total of 36 teeth and geldings and stallions 40 teeth.

Occasionally 'wolf-teeth' may erupt in the upper jaw. These are vestigial pre-molar teeth which may cause problems, interfering with the bit action and causing pain. It is probably best to have them removed by the horse dentist or vet.

The horse's teeth are designed to eat grass; the incisors can nibble short grass close to the roots while the ridged surface of the molars grind the tough fibrous grass before swallowing.

The horse's head is shaped so that the upper jaw is wider than the lower one – in other words, the molars overlap each other at the sides. Horses' teeth grow continually upwards out of the gum and are continually worn away by grinding fibrous food. As the jaw moves from side to side during chewing the molars grind against each other, but the lower ones do not reach the outer edge of the upper molars and the upper molars never reach the inner edges of the lower molars. Eventually these edges become long and sharp and can cut the cheeks and tongue, making eating and carrying a bit painful. Quidding, which is the dropping of half-chewed food out of the mouth, is a sign of sharp molar teeth.

The vet or horse dentist can file the edges of the molar teeth, levelling them with the rest of the molar teeth, a process called rasping or floating. To avoid problems, a horse's teeth should be checked twice a year and rasped if necessary.

Summary of Part II

The Effects of Exercise on the Body

As the horse begins to work, there is an increase in muscular activity. Muscles need energy to contract; initially, this energy is supplied by the breakdown of fuels stored within the muscle cells. When these stores run out, fuels are supplied from other part of the body, e.g. the liver, and brought to the muscle cells in the bloodstream in the form of glucose and free fatty acids.

Releasing energy for muscle contraction from these fuels with optimum efficiency requires oxygen. Oxygen is supplied by the lungs via the red blood cells which carry it in chemical combination with haemoglobin. A toxic waste product, carbon dioxide, results from this fuel breakdown, and it must be removed from the cells if the muscles are to continue working. The carbon dioxide is removed, attached to haemoglobin, by the bloodstream and expelled from the lungs.

This means that not only does the bloodstream supply oxygen and other nutrients, but it also removes waste products. The harder the horse works, the more rapidly its muscles contract and the more oxygen is needed. The heart, which is responsible for pumping blood round the body, must increase its rate and power of pumping as exercise increases. The rate and depth of breathing must increase so that more oxygen is drawn into the body. Together, the increased heart and respiration rates ensure an increased supply of oxygen to the muscle cells.

As exercise increases, the adrenal glands release the hormone adrenalin. This promotes a fight-or-flight reaction in the horse, increasing both breathing and heart rates, and mobilising body fuels from their storage depots. The spleen also contracts, thus releasing a large volume of blood into the circulation and increasing the oxygen-carrying capacity of the body, aiding muscle contraction.

During very strenuous exercise, when the muscles are working (i.e. contracting) at maximum capacity, the oxygen demands of these muscles exceeds the amount of oxygen that can be supplied by the blood. The energy for contraction has to be supplied from fuel supplies in the absence of oxygen, or anaerobically. This anaerobic respiration is not as efficient in producing energy as aerobic respiration, and can be described as high short-term power at the expense of limited fuel sources. A poisonous waste product, lactic acid, is also produced, and is considered to be a major contribution to muscle fatigue.

Lactic acid is taken in the blood to the liver, where it is changed to non-toxic substances in the presence of oxygen, i.e. at the end of anaerobic exercise. Excessively high levels of lactic acid are associated with muscle cramps and muscle damage. At the end of exercise, a horse blows until all the lactic acid is removed from the blood. At slower levels of exercise, oxygen demand equals oxygen supply and lactic acid does not accumulate.

Part III

The Performance Horse – Problems and Care

9 Lameness

Tendon Injury

Tendon Structure

Figures 9.1 and 9.2 show the anatomy of the horse's lower limb. Tendons consist of bundles of individual collagen fibres on which the strength of the tendon depends. These collagen fibres are longitudinally orientated and interspersed with fibroblast cells. The fibroblasts produce a material which, under certain conditions, joins together to produce long collagen fibres. The conditions needed are correct salinity, acidity and oxygen supply.

Collagen production in the tendon is continuous, which means that tendons are dynamic structures like bones. Normally, the collagen in the tendons of the lower limb is entirely renewed in six months. Special enzymes trim off excess collagen and also dispose of damaged and old fibres. This constant formation of collagen means that it is very important that there is satisfactory blood flow through tendon tissue, bringing oxygen and nutrients. However, the blood system through the tendon is very modest and easily upset, even by very sligh strains and sprains, reducing the oxygen supply and leading to degenerative changes

Blood Supply

Each tendon is covered by a thin layer of connective tissue, which continues into the fibres and separates bundles of fibres; this tissue also carries blood, lymph and nerves. The whole tendon is usually enclosed in the tendon sheath.

There are 'cold regions' in the horse's lower limb where the blood supply is less efficient. These are:

- check ligament
- middle of the superficial digital flexor tendon
- around the fetlock joint

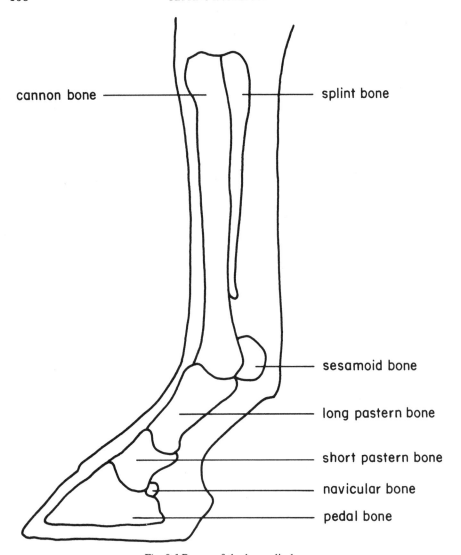

cannon bone —————————————— splint bone

—————————————— sesamoid bone

—————————————— long pastern bone

—————————————— short pastern bone

—————————————— navicular bone

—————————————— pedal bone

Fig. 9.1 Bones of the lower limb

It is in these regions that degenerative changes commonly occur; they are seen as slight heat and swelling. Any heat in the lower limb should be taken seriously, as ignoring these early warning signs may lead to more severe problems later on.

Collagen Fibre Structure

Collagen is the most abundant protein found in the mammal's body, being present in skin, bone, tendons and ligaments. There are four types of collagen:

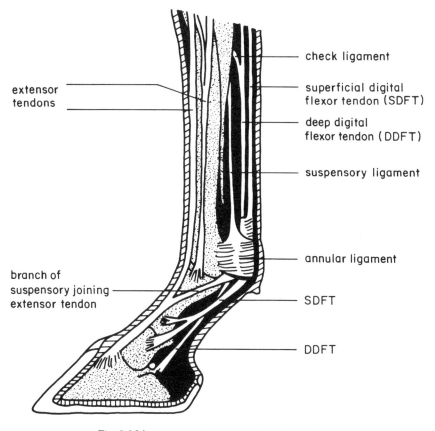

extensor tendons

check ligament

superficial digital flexor tendon (SDFT)

deep digital flexor tendon (DDFT)

suspensory ligament

annular ligament

branch of suspensory joining extensor tendon

SDFT

DDFT

Fig. 9.2 Ligaments and tendons of the lower limb

Type I: bone and tendon
Type II: cartilage
Type III: foetal membrane, cardiovascular system
Type IV: basement membrane

Collagen type I is the strongest form, the strength arising from its 'crimped' structure. Collagen itself is inelastic, but if put under strain the crimped structure can be pulled straight, allowing the fibre to increase its length by 4% and thus absorbing strain.

As the collagen fibres age, or as they are used, the amount of crimp decreases; the tendons become less efficient and are replaced. Collagen fibres are replaced in response to the demands made upon them, so in the working horse fibres are replaced quickly; the average age of the collagen fibres is younger and consequently the amount of crimp is greater, and the amount of stress and strain that can be tolerated greater. In the idle horse the turnover of collagen decreases and crimp is lost and not replaced, so if such a horse is brought into work suddenly the tendons are less likely to withstand the work load.

Getting Tendons Fit

Traditional fittening programmes involve weeks of walking work to 'harden the horse's legs' and while the terminology may be confusing, there are sound practical and physiological reasons for beginning the programme with gradually increasing amounts of walking. The unfit horse has tendons containing elderly collagen fibres, lacking in crimp and thus unable to stand faster work without the risk of damage. Walking work stimulates the blood flow to the tendons and encourages increased turnover of collagen fibres, resulting in stronger, more efficient tendons by the time faster paces are introduced into the training programme.

Injury

The horse appears to be particularly susceptible to tendon problems of the lower limb, and tendon injuries are a major cause of lameness in all performance horses. Tendon injury falls into two broad categories:

- mechanical
- degenerative

Mechanical damage is the result of injury – for example, a horse striking itself or being struck by a sharp or blunt object. Wounds due to glass or wire can occur, most commonly below or on the fetlock. The nature of the wound may vary from a bruised to a cut or severed tendon. When the tendon is bruised, there is haematoma; the fibres in the tendon become separated by bleeding within the tendon, causing painful swelling. This haematoma will eventually be reabsorbed and the swelling will subside. However, there may be scarring within the tendon and adhesions may form. Adhesions are areas where healing tissue becomes 'stuck' to overlying and underlying tissue, thus limiting the freedom of movement of the part. Mechanical damage can be particularly serious if the wound becomes infected. Recovery is slow, and there may be extensive scarring and adhesion. The superficially situated tendons, the superficial digital flexor tendon and the extensor tendons (see fig. 9.2), are the most susceptible to mechanical damage. Sudden strains or sprains may occur after a specific incident, such as putting a foot in a hole, overtiring an animal or running into a very wet piece of ground.

Degenerative damage is more complex, and is essentially provoked by a lack of oxygen in the tendon leading to partial or complete tendon

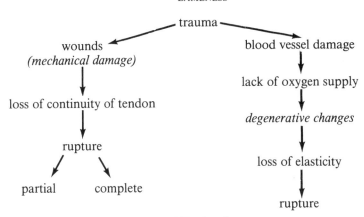

Fig. 9.3 Tendon damage

rupture. The onset of 'a leg' is more insidious, arising from repeated stress (fig. 9.3).

Cases of Tendon Injury

(1) *Muscle fatigue.* When the muscles are tired, muscle co-ordination can become upset; the muscles to which the tendon is attached may become inelastic, so that all the elasticity falls on the tendon, stretching it beyond its safety margin. Muscle fatigue will occur more rapidly in heavy and uneven going, and if the horse is jumping.

(2) *Fitness.* Unfit horses with little muscular strength become tired and unco-ordinated more easily.

(3) *Foot care.* Bad or irregular shoeing can lead to the horse having toes that are too long and heels too low, causing overextension of the fetlock joint and strain on the flexor tendons.

(4) *Conformation.* Horses that are too heavy for their limbs, have long cannon bones, long pasterns, crooked limbs, etc. are all more at risk due to the extra strain being placed on the tendons.

The Signs of Tendon Injury

Mechanical Injury

The signs of mechanical injury are usually dramatic, but vary according to which tendon is damaged and the extent of the injury. The horse will be lame and, even though the exposed or frayed tendon may be seen within the wound, there will be little bleeding: veterinary attention will be necessary. These wounds always cause great concern, but generally mechanical injury is less of a healing problem than tendon damage resulting from degenerative change.

Degenerative change

Degenerative change can be an insidious process, although strains and sprains can happen quickly, e.g. if a horse gallops into very soft, boggy ground. Several hours after exercise the tendon will show signs of inflammation, i.e. heat, swelling and soreness over the tendon. One or both forelegs may be affected, and either or both tendons may be involved. In acute sprains, the horse may pull up lame or go lame shortly after the injury. The horse may stand with the heel held up to ease pressure on the flexor tendon area. The knee will be cocked forward at rest and the horse will not allow the fetlock to drop when moving to avoid painful pressure. The tendon may have a 'bowed' appearance (fig. 9.4).

The Effects of Tendon Injury

The initial tendon injury is due to tearing of some of the tendon fibres and haemorrhage (bleeding) within the tendon leading to inflammation. The tendon is surrounded by firm tunnels and encircling ligaments (annular ligaments), and it is this that causes a great deal of the damage in a bowed tendon. The ligaments cause pressure and damage when the injured tendon swells, leading to tissue death which may involve the tendon sheath. Dead tissue is replaced by scar tissue, which is inelastic and laid down in a haphazard fashion.

Thus it is the inflammation that causes much of the damage which leads to effects such as the tendon sheath telescoping around the deep and superficial flexor tendons, and adhesions being formed between the

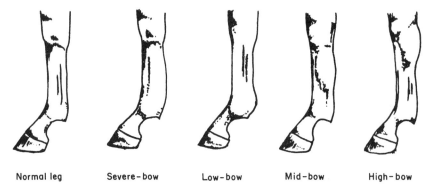

| Normal leg | Severe-bow | Low-bow | Mid-bow | High-bow |

Fig. 9.4 Different areas of tendon strain

tendons and tendon sheath. This means that the tendon loses its ability to stretch and is much more susceptible to re-tearing.

Healing of Tendons

Healing is preceded by an inflammatory process involving heat, swelling and pain. The tendon fibres are damaged and the body's reaction is to cause cells to move into the damaged area. Here, these cells and others released from the damaged blood vessels dissolve away the damaged fibres, blood clots and debris associated with the wound. The damaged fibres are replaced with granulation tissue, which is a repair tissue with a large number of cells rather than fibres, leading to the tendon being permanently thickened.

Gradually, scar tissue containing collagen fibres is laid down. However, these collagen fibres are laid down in a haphazard fashion, not longitudinally in the direction of pull along the tendon, reducing the strength of the tendon. In time these fibres realign themselves so that after six months the tendon looks more like normal. Also, the new collagen laid down is type III collagen, not type I collagen; type III collagen lacks crimp and is a weaker, immature form less able to withstand stress. Thus apparently 'healed' tendons will have a greater susceptibility to further injury. Type III collagen is eventually replaced by the original type I collagen, but this may not occur until more than twelve months after the injury.

It is important that the new collagen fibres align themselves longitudinally, rather than at random, to give the tendon strength. This longitudinal arrangement depends on force being applied to the healing tendon. Force is generated by short periods of walking in hand once the initial heat and pain has subsided. Exercise also helps minimise adhesions between the collagen fibres and the tendon sheath.

Treating Tendon Injury

The first step in treating an injured tendon is to control the inflammatory reaction. This will reduce the amount of scar tissue that is formed in the tendon. Once the pain and swelling are reduced, various treatments can be given including, rest, firing, tendon-splitting and carbon fibre implants. Prompt attention can mean the difference between returning to work or retiring after a strained tendon.

Pressure bandages, hot and cold compresses and ultrasound physiotherapy can be used. Pressure bandaging is not primarily for support but to cut down the inflammation in the tendon area. Anti-inflammatory drugs can be given and a plaster cast may be applied by the veterinary surgeon. This will prevent further swelling and provide support for the leg, and will stay on for seven to ten days. A second cast may be needed for two to three weeks, depending on the severity of the injury. After the second plaster has been removed, light exercise in hand will help prevent the formation of scar tissue adhesions in the tendon before the horse is turned away for nine to twelve months.

Controlling the Inflammatory Reaction

Inflammation is part of the healing process, but if the leg is excessively swollen more harm than good may be done, and healing may be delayed.

Cold water treatments (e.g. cold bandages, hosing (fig. 9.5), freeze packs) decrease swelling and pain temporarily, and are followed immediately by support bandages to stop excess flow of blood back to the leg. Bandages must be carefully applied to wet legs as they may shrink, and *both* front legs should be bandaged as the sound leg will be taking

Fig. 9.5 Cold hosing the lower limb

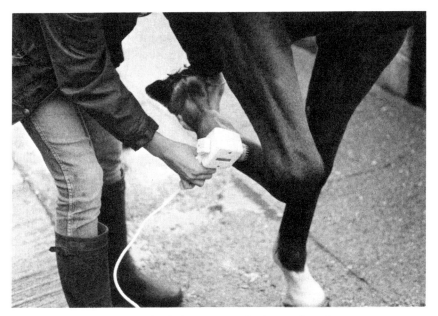

Fig. 9.6 Mechanical treatment of a damaged tendon

most of the weight and so needs support.

Heat therapy (e.g. poultices, ultrasound, faradic, magnetic therapy) increase blood supply to the area helping to dispel swelling. Heat treatments should not be used for 48 hours after the initial injury, when the pain has been minimised by cold treatments.

Alternate hot and cold therapy acts as a kind of massage, reducing then stimulating flow of blood to the injury.

Massage (e.g. hand rubbing or mechanical vibration – fig. 9.6) will decrease swelling and help avoid scar-tissue adhesions. Exercise is a cheap form of therapy and the best way of dispelling swelling of the horse's lower limb, provided it is controlled and used with judgement. Once the initial lameness has reduced so that the horse can walk comfortably, five minutes walking in hand two or three times a day can be very helpful.

Further Tendon Treatments

Counter-irritation treatments promote the inflammatory reaction by introducing additional irritation. They vary in their severity, the mildest being liniments and the most severe firing.

Rubifacient drugs (e.g. liniments, tightners, bracers) are mild applications, accompanied by massage and bandaging, which have a very temporary warming effect and are therefore applied two or three times a day. Blisters are more dramatic in their effect and vary from mild, working blisters causing red, inflamed skin to severe blisters which cause peeling of the skin. Their effect is very superficial, affecting only the skin and immediately underlying tissue, so there is no justification in their use for treating tendons.

Firing, the application of hot iron or acid to the horse's legs, has been shown to be at least ineffective and at worst detrimental to tendon healing. The scar tissue produced, far from providing a support to the leg, is both thinner and weaker than normal. The only possible advantage of firing is that it enforces the required rest. The pain it produces then subsides gradually, producing an effect of gradually increasing mobility. Time is the best healer in the case of tendon injury, however, and patient self-discipline on the owner's part is less painful for the horse, less costly and more effective.

In the tendon splitting operation a scalpel is used to make vertical slits through both the damaged tendon and the surrounding healthy area, the aim being to allow the healing elements to penetrate right into the damaged area. However, recent work has shown that tendon splitting amounts to no more than a sophisticated form of firing and actually delays healing.

The relatively new technique of carbon fibre implants is becoming increasingly popular in the treatment of acute tendon injuries, including ruptured and severed tendons, as well as for the treatment of old, chronic injuries. It involves the use of a plait of carbon fibre filaments which is inserted into the damaged tendon, acting as a scaffolding along which tendon tissue is produced with the fibres running in the direction of the tendon, not at random as in scar tissue.

As with all tendon treatments, carbon fibre implants can only be effective if the horse is given adequate rest after the operation – at least nine months – and then brought back to work slowly and carefully.

Corrective shoeing can help in the treatment of tendon strain. If the suspensory ligament is not involved, the heel can be raised 25 to 40 mm (1 to 1.5 in) for four to six weeks to take the weight off the bowed tendon. This should be lowered gradually over a period of six to eight weeks. The horse may benefit from being shod with pads or wedges when brought back into work.

Treatment of Chronic Tendon Injuries

Chronic tendon injuries result from neglected acute tendon inflammation or years of unnoticed slight heat and swelling that gradually builds up into a firm, prominent swelling on the back of the cannon. This swelling is actually scar tissue in the tendon caused by inflammation at the time of injury.

Tendon injuries are a common cause of lameness in the competition horse and, while it is accepted that they can occur accidentally and unexpectedly, they are far less likely to occur in a correctly trained, fit horse. A horse should not be expected to do any fast work before its legs and muscles are ready, and tired horses should not be pushed beyond their capabilities. Any signs of heat and swelling in the lower leg should be treated with great care.

Once a horse has sustained a tendon injury, it will always need careful management as the recovered tendon is unlikely to be as strong as one that has never been injured. Although better shoeing and improved training methods may reduce the incidence of tendon damage, and techniques like heat sensitive cameras and force plate studies lead to earlier detection of lameness, damage is certain to occur when horses are pushed to the limits of their performance.

Ligament Injury

The suspensory and check ligaments can also suffer strains and rupture. Injuries occuring to the suspensory ligament at its upper attachment to the cannon bone are very serious. The fine fibres of the ligament fit into tiny holes on the surface of the bone. Once torn loose, the scar tissue formed is too bulky to allow them to return into the holes and instead they adhere to the bone surface, this weak attachment easily loosens.

Signs of Suspensory Ligament Injury

Lameness and swelling over the suspensory area are signs of injury to the suspensory ligament. If the leg is lifted and the relaxed suspensory palpated, the horse will show signs of pain. The horse will hold the knee and fetlock forward and raise the heel slightly off the ground; the fetlock will not be lowered all the way when the horse is moving.

Treatment of Suspensory Injury

Acute strains require complete rest, anti-inflammatory treatment and ultrasound, if available.

Chronic strains may be treated with corticosteroids and rest – at least six months. Blistering and firing have been used in the past with doubtful results.

In all types of suspensory ligament injury, a considerable amount of time must be allowed for healing because of the poor blood supply to the ligament.

Navicular Disease

Previously considered incurable, progress is being made in the treatment of navicular disease, a nightmare of all horse-owners.

The Position of the Navicular Bone

The elongated *navicular bone* is a sesamoid bone lying transversely inside the hoof above the middle of the frog between the *short pastern* and the *pedal bones* (fig. 9.7) and held in position by ligaments. The navicular bone articulates with the short pastern bone, and during movement the weight of the horse presses down and pushes the navicular bone down and back. The back and lower borders of the bone are in close contact with the *deep digital flexor tendon* which rides over the navicular bone before attaching to the pedal bone. In order to allow the tendon to run smoothly over the bone, the back border of the navicular bone is covered in cartilage and a pocket of lubricating joint fluid, the navicular bursa. This area is most under pressure during movement, especially when the foot is on the ground and the weight of the horse is travelling over the leg.

Diagnosis of Navicular Disease

The onset of navicular disease is usually gradual and includes signs such as loss of action, shortening of the stride or reluctance to jump. The lameness may be intermittent, the horse coming sound after rest but being lame again if worked. The horse may only be slightly lame, and one or

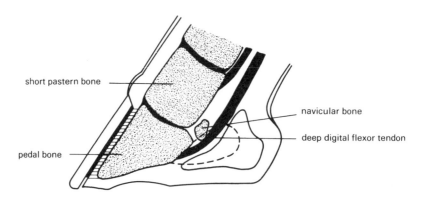

Fig. 9.7 The position of the navicular bone

both front feet may be affected. Alternatively, the onset may be sudden – after hard work, for example – and the horse may be lame in only one leg.

As the disease progresses, the horse becomes increasingly lame and avoids putting weight on the heels of the foot, so that the toe lands first. This causes the shuffling gait, worn toes and contracted heels character-istic of a horse with navicular disease.

Positive diagnosis of navicular disease is often difficult. Initially, the lameness must be identified as being in the foot, and then more specific tests for navicular must be carried out. Standing the horse with its toe on a block of wood, e.g. the shaft of a hammer, with its heels unsupported, stretches the flexor tendon which in turn presses on the navicular bone so that the horse is more lame when trotted up. Pressure to the middle of the frog with hoof testers will increase lameness. The posterior branches of the nerves to the foot can be blocked with local anaesthetic so that pain is removed from the back of the foot. If after this the horse goes sound, navicular disease would be strongly suspected.

The next step is to X-ray the foot. However, there can be confusion with interpreting X-rays. In a positive diagnosis, one would expect to see widening of the blood vessels supplying the navicular bone; this bone will open out to form an inverted vee-shape as bone is reabsorbed along the line of the blood vessels, giving characteristic 'tree-shaped' erosions of the navicular bone. Some vets say that up to five of these pyramids are normal.

Erosion of the Navicular Bone and Cartilage

Navicular disease causes erosion of the cartilage on the back border of the navicular bone and spurs of new bone on the side edges; there may also be changes in the flexor tendon running over the navicular, including bruising and discolouration, eventually causing adhesion between the tendon and the bone. In extreme cases, there may be fracture of the weakened navicular bone or rupture of the tendon.

The erosion of the cartilage and bone can be detected by X-ray as areas of translucency. New bone formation can be seen, and also enlarged openings for the blood vessels supplying the bone as a result of inflammation in the bone. The erosion of bone and cartilage appears to be due to arteriosclerosis of the blood vessels supplying the bone. In other words, the tiny blood vessels are blocked by blood clots so that areas of bone are deprived of the blood supply which brings oxygen and other nutrients.

This condition is called ischaemia, and is associated with pain and the death of small areas of bone. The pressure of the flexor tendon wears away these areas of weakened bone, causing the erosion seen on X-ray. The tendon will now be running over an abrasive surface and this causes bruising. New blood vessels also develop in an attempt to supply the bone deprived of blood and nutrients, causing mushroom-shaped canals in the lower border of the bone.

However, horses very lame with suspected navicular may show no signs on X-ray while sound horses may have lesions, leading to some confusion in interpreting X-rays. A true diagnosis may be obtained from repeated X-ray examination at four to six-month intervals to monitor the progression of the changes in the bone.

Treatment

As the cause of navicular disease is not fully understood, treatment is aimed at alleviating the symptoms rather than effecting a cure. The following are possible causes:

(1) Continual compression of the navicular bone by the deep digital flexor tendon.
(2) Repeated concussion of the navicular bone from below by the frog.
(3) There may be a hereditary disposition, possibly related to conformation, e.g. boxy feet, upright pasterns, and an incorrect foot and pastern axis. Boxy feet may, however, be a result of the disease and not the cause.

Navicular disease occurs in all types of horses, although those affected do tend to be middle aged and to have been subjected to fairly hard work.

The farrier may attempt to open out the boxy, navicular foot by grooving the hoof, cutting the heels down and using surgical shoes. An alternative is to allow the heel to grow to relax pressure on the flexor tendon. Bar shoes, pads and wedges will also have this effect.

Phenylbutazone ('bute') will only alleviate the symptoms, allowing the horse to work when there are no other alternatives, and as such should not be abused.

Warfarin, an anti-clotting drug, is given in the feed, the dose being adjusted to increase clotting time by 20 to 50% depending on the severity of the lameness. Repeated and costly blood sampling is necessary to monitor clotting time as fatal haemorrhages may occur; however, this treatment is well established.

Isoxsuprine hydrochloride is a relatively new treatment, the effect being to dilate (widen) the blood vessels supplying the navicular bone. A high degree of success has been claimed, particularly if early diagnosis and immediate therapy is given.

A common operation which may be carried out in cases of navicular disease involves cutting part of the nerve supply to the foot so that sensation is lost in the posterior part of the foot. Although generally considered a last resort, this will enable a chronically lame horse to continue work. However, very careful attention must be paid to the foot and shoeing.

A new operation aimed at reducing pressure on the navicular bone is still at the early trials stage.

10   Metabolic Disorders

Azoturia and the Tying-Up Syndrome

Azoturia is known by many names – setfast, Monday morning disease, tying-up, paralytic myoglobinuria and blackwater – and it is a surprisingly common problem which can in some cases be very severe. Traditionally, it occurs in the fit, stabled horse which has had an enforced rest, e.g. bad weather conditions or the Sunday rest day, while remaining on its full corn ration; it manifests itself on the horse's return to work.

Symptoms

When the horse is exercised it starts work normally, but after a while begins to feel stiff behind or after a halt may be unwilling to go forward, it may blow and/or sweat excessively for the type of work being done. If the animal were forced to continue, it would eventually be unable to stand and would go down. It will be sweating profusely, breathing rapidly and showing evidence of acute pain. The muscles of the loins and hindquarters may be swollen and will certainly be tense and painful. Any urine eventually passed will be discoloured, the actual colour varying between amber, port and almost black – hence the name 'blackwater'.

   Azoturia can affect all breeds of horses, all ages and both sexes, although it does appear to be more common in mares, Quarterhorses and Draught breeds. If a horse has had azoturia once, it is likely to recur, and such horses must be carefully managed. It is usually exercise-related, although exceptions may occur –azoturia can, for example, follow a bout of colic. The condition affects skeletal muscle, usually the back muscles on both sides; however, occasionally forelimbs can be affected. In all cases, the horse will be in pain and the gait will be affected, varying in degree from slight stiffness to an inability to stand.

Treatment

Very mild cases showing slight stiffness may wear off with gentle walking exercise. This form is generally referred to as 'tying-up' and many horse-owners do not recognise it as a mild case of azoturia.

The majority of cases require complete rest – the horse should not even be walked home. A coat should be thrown over its loins to keep them warm, and the animal boxed home. Even the muscular exertion required for travelling may cause further damage. Once in its stable, it is essential to keep the horse warm, particularly over the back region. All concentrate feed should be cut out of the ration and only a minimum of hay and a laxative bran mash given.

The vet should be called for all but the mildest cases. Treatments may include several of the following: phenylbutazone or corticosteroids to help reduce inflammation in the damaged muscles, a muscle relaxant, a sedative, and vitamin E with selenium. Recumbent cases may need intravenous fluid and intensive nursing.

The vet may also take a blood sample and test for the presence and concentration of certain muscle enzymes (C.P.K., A.S.T. and L.D.H.) which can indicate the severity of the attack. Further blood tests will help monitor recovery. C.P.K. is released after stress or damage to muscle. The concentration reaches a maximum in the blood six hours after the muscle damage has been sustained and gradually returns to normal, provided there is no further muscle damage. However, C.P.K. levels can *normally* increase by five times in an unfit horse made to work, so interpretation of readings can be difficult. A few typical clinical cases may not show any increase in C.P.K. levels. The reason for this is not known.

The period of rest necessary may vary from a few days to weeks. Exercise must be introduced very gradually, the first day's exercise consisting of 5 minutes walking in-hand, building up periods by 5 to 10 minutes per day. A sharp look-out should be kept for any recurrences of the symptoms. There may be some scar tissue replacing damaged muscle tissue, and hence some muscle wastage, although this may not be noticeable.

Cause

The cause of azoturia is not really known but it is assumed that there are many contributing factors, including exercise of an unfit horse, irregular exercise and rest on a normal ration. All of these may lead to an upset in carbohydrate or fat metabolism. A deficiency of vitamin E, selenium or electrolytes and an excess of certain types of food, e.g. barley and flaked maize, have also been implicated in the condition.

An upset in normal carbohydrate or fat metabolism can cause a build-up of lactic acid in the muscles, which causes muscle damage. The muscle pigment myoglobin is released from the damaged cells and excreted by the kidney via the urine. The myoglobin accounts for the discolouration of the urine of an affected horse. However, the kidney is not designed to cope with excreting myoglobin, and kidney damage can result from severe cases.

Precautions

Precautions can be taken to avoid azoturia:

(1) Feed a correct, balanced diet.
(2) Feed according to work.
(3) Keep feed ahead of work, i.e. anticipate a rest day by cutting the corn ration. If this is not possible, turn the horse out or walk it out in-hand for 10 minutes.
(4) Gradually increase the horse's work load, allowing the muscles to adapt to exercise.
(5) Prime the system by thoroughly warming-up the horse before any hard work is done.
(6) Ensure that the horse – especially one that is fit and on a high concentrate ration – is kept in regular work.

Laminitis

Laminitis is commonly known as a problem of fat ponies, particularly in the spring. While such ponies are very much at risk, so also are moderately worked horses who are routinely fed a high proportion of corn in their daily ration. Many horses walk a line just this side of a laminitis attack, and it may be responsible for some bouts of sudden, temporary lameness.

Symptoms

In an acute attack, the feet are usually but not always hot and the horse shows evidence of pain, i.e. sweating and blowing. The stance with the front feet extended, weight taken on the heels and the hind feet crouched underneath to bear the weight is characteristic of laminitis (fig. 10.1). A useful symptom to look for is a full throbbing pulse in the digital artery.

The symptoms may not be so dramatic in a slower-developing case of laminitis, early signs being withdrawn behaviour, off-feed, reluctance to move and a general air of discomfort. However, in an animal prone to laminitis immediate action should be taken.

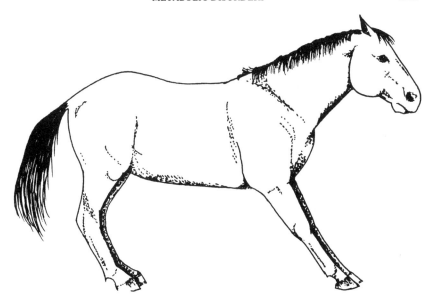

Fig. 10.1 Typical stance of the laminitic horse

Causes

(1) *Digestive*. Grain overload, spring grass, cold water in the stomach of a hot horse.
(2) *Postparturient*. Metritis, perhaps due to retention of some after-birth or a very arduous foaling.
(3) *Mechanical*. Direct damage to the sensitive hoof due to concussion, or excessive foot trimming and rasping, e.g. before height measurement.
(4) *Stress*. Illness, long journeys, or a change in environment.
(5) *Drugs*. Steroids, wormers and antibiotics may cause a reaction in sensitive horses.

Laminitis is not a disease of the hoof but of the circulatory system. The causes listed above result in the release of endotoxine (poisons produced in the body). These toxins are vasoactive – they cause the tiny capillaries in the hoof to contract, thus denying the sensitive structures of the foot nourishment. Blood continues to travel to and from the foot but it fails to reach toe cells holding the sensitive and insensitive laminae together. The undernourished laminae weaken and die, freeing the pedal bone from some of its essential attachments so that the unopposed pull of the deep flexor tendon causes it to rotate and the sole of the foot drops.

Treatment

(1) Remove the cause, e.g. the vet may administer mineral oil by stomach tube to flush out the grain from the system, or he may remove retained afterbirth. Reduce the diet.

(2) Forced exercise will help improve circulation. The floor should be soft and supportive to the sole, e.g. sand, peat or muck.

(3) The vet may administer antihistamine, anti-inflammatory and pain-relieving drugs, e.g. corticosteroids and phenylbutazone to counteract the changes taking place in the horse's body.

(4) Severe cases where the pedal bone has dropped will need remedial foot trimming to realign the bone, and surgical shoeing may also be helpful so that the horse can work normally.

Anaemia

Anaemia is not a disease but a condition of the blood. An anaemic horse is one which has an abnormally low level of haemoglobin, the oxygen-carrying substance contained in red blood cells, in its blood.

Symptoms

The general signs of anaemia are poor coat, pale membranes of eye and gum, muscular weakness, depression, lack of appetite and an increased rate and force of heart beat. The normal red blood cell count is 7 to 8 million per cubic millimetre and the normal haemoglobin levels are 11 to 17 g per 100 ml of blood. Figures indicating anaemia would be 3 to 6 million red blood cells per cubic millimetre and 8 to 11 g haemoglobin per 100 ml of blood. Values such as these reduce the oxygen-carrying capacity of the blood, having an adverse effect on the horse's ability to work.

Causes

Anaemia can be due to:

(1) Loss of blood as a result of rupture of a blood vessel, external or internal bleeding, i.e. haemorrhage.

(2) Increased destruction of red blood cells.

(3) Nutritional deficiency.

A temporary state of anaemia can be caused by severe or recurring bleeding following serious wounds or episodes of epistaxis (nosebleed). These anaemias are easily reversed if the overall body condition is not lowered by infection or continued blood loss. Bleeding involves the loss of all components of the blood: red and white blood cells and plasma and transfusions of plasma or whole blood may be required if bleeding is severe.

Destruction of red blood cells is called haemolytic anaemia, and can be due to infection with bacteria, virus, protozoa, poisoning or immunological reaction.

Poor-quality feed can cause the bone marrow to cut the output of red blood cells and haemoglobin. Normal production and replacement of red blood cells depends heavily on good nutrition and proper health. The most common causes of a low red blood cell count are a heavy redworm (*Strongylus vulgaris*) infestation and poor-quality feed. The lack of copper, iron, folic acid and vitamin E will result in red blood cells deficient in haemoglobin.

Treatment

Anaemia can only be alleviated by treating its source first. Poor nutrition, parasite burdens, viral and bacterial infections and haemorrhage must be dealt with before the anaemia can be treated. The removal of the source may actually take enough pressure off the body to allow it to correct the anaemia spontaneously within a month or two. For immediate treatment, a supplement containing iron and B-group vitamins can be fed, and injections of vitamin B_{12} act as a haemoglobin booster.

Anaemia can usually be prevented by good stable management, including a regular worming programme and a good-quality, balanced diet.

11 Respiratory Problems

Obstructive Pulmonary Disease (O.P.D.)

Breathing is essential for life, and any defect in the horse's respiratory system can dramatically decrease performance, especially in horses working at maximum effort. It has been long recognised that horses living in dusty stables cough, and if forced to continue to work in these conditions eventually become 'broken-winded'.

Causes and Symptoms

Research work has shown that some horses are sensitive to the spores of fungi found in large numbers in badly conserved hay and straw, notably *Aspergillus fumigatus* and *Micropolyspora faeni*. When these spores are inhaled, they stimulate a normal defence mechanism, i.e. the bronchi and bronchioles leading to the lungs constrict in an attempt to stop the particles penetrating any further into the lungs. This is usually accompanied by a cough, causing expulsion of the spores, and then the bronchi return to their normal diameter.

However, in horses that have become sensitised to fungal spores the bronchioles fail to return to normal and substances released by the body cause an inflammatory reaction, swelling (oedema) and increased mucus production, all of which serve to reduce the diameter of the airways in the lung. This is known as 'small airway obstruction' or S.A.O. (fig. 11.1) when the airways are narrowed, there is increased resistance to airflow and breathing out takes longer; the next inhalation may have started before all the previous breath has been expelled. This means that air is trapped in the alveoli which overinflate, a condition characteristic of broken wind (emphysema).

The consequences of these effects on the horse can be very serious. Instead of the respiratory movements being barely noticeable, the resting horse has to make a physical effort to expel the inhaled air. This results in a wheeze, a double exhalatory effort and eventually the development of the 'heaves' line, which runs between the end of the ribs and the belly (fig.

11.2) and is due to the overdevelopment of the muscles controlling respiration – hence the old name 'heaves' for this condition. The horse will also develop a chronic harsh cough and have an increased respiration rate. Exercise tolerance, i.e. the horse's performance, may be dramatically reduced.

Stable dust also contains micro-organisms, which can cause inflamma-

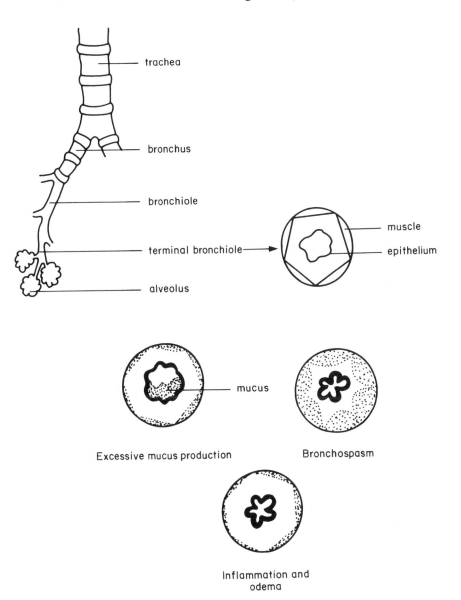

trachea

bronchus

bronchiole

terminal bronchiole

alveolus

muscle

epithelium

mucus

Excessive mucus production

Bronchospasm

Inflammation and odema

Fig. 11.1 Normal and obstructed small air passages of the lung

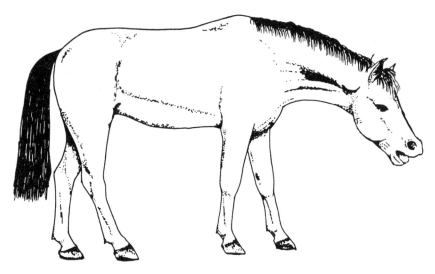

Fig. 11.2 The 'heaves' line

tion of the bronchi (bronchitis) and bronchioles (bronchiolitis), and fine dust particles which irritate and damage the lining of the tubes, causing coughing.

Treatments

Fortunately, obstructive pulmonary disease is reversible if treated promptly, and with the correct combination of stable management and durg therapy the horse can return to full work.

The stable management must be such that the horse lives in as dust-free an environment as possible. This involves three aspects:

- ventilation
- bedding
- feeding

A dry, clean, airy stable is the first consideration. The ventilation should be such that there are six complete air changes every hour, which means that there should be inlet and high-level outlet air vents other than the door. Stables must be kept free from dust and cobwebs, and preferably should not be down-wind of dusty areas such as barns.

Straw is not an acceptable bedding for a horse afflicted with O.P.D. and alternatives such as shredded paper, peat or shavings should be used. All wet bedding must be removed, because after several days urine-soaked paper can become an ideal medium for fungal growth and thus the production of spores. Deep-litter bedding would not be suitable for an O.P.D. horse.

Complete cubes and haylage provide spore-free alternatives to hay. Both are expensive ways of feeding, and cubes can lead to boredom and problems due to lack of roughage in the diet. Haylage must be stored carefully and used within three days of opening. Soaking hay is a successful, if time-consuming, way of feeding the allergic horse, provided it is good-quality hay – it should not be used as a way of making poor-quality hay palatable. Veterinary opinion varies as to whether the hay should be soaked for 24 hours or if a 10-minute thorough dunking is adequate. A working compromise would be to put the following morning's hay in to soak at evening stables and the night-time hay in soak in the morning, i.e. 8 to 12 hours soaking.

Hay and straw carried past the affected horse's stable or stored in close proximity can start symptoms, so ideally the O.P.D. horse should be stabled separately.

Drug Therapy

Occasionally the horse will be exposed to spores, e.g. at a show or an event. Here, drug therapy becomes useful. Three main groups of drugs are used:

- mucolytics
- antispasmodics
- anti-allergy

Mucolytics, e.g. Bisolvon, decrease the viscosity of the mucus produced by the bronchioles, making it easier to expel the mucus and clear the airways.

Antispasmodics, e.g. Ventipulmin, act on the constricted bronchioles and relax them, thus allowing a free passage of air in and out of the lungs.

Anti-allergics, e.g. Chromovet, act on the cells involved in the allergic reaction, desensitising them for up to 10 days.

Using these drugs and proper stable management, the O.P.D. horse should be able to perform to high levels. However, it is possible that there are a vast number of horses suffering from subclinical symptoms, and it may be worth bedding on shavings and soaking hay for all performance horses in order to avoid such problem from the outset.

Respiratory Viruses

Viruses are very small living particles which infect and cause disease in man, animals and plants. Unlike bacteria, they are not killed by antibiotics. There are three major viruses causing respiratory disease in horses:

- influenza
- equine herpes virus 1 (rhinopneumonitis)
- adenovirus

Typically, these viruses infect a number of horses at the same time. Many older horses have experienced the disease and are resistant to reinfection so respiratory disease primarily effects the young horse, making 'the virus' of particular importance in racing stables. However, if a mutant (genetically changed) version of the virus appears it will affect all horses, because even the older horses will have no resistance to it.

The symptoms caused by the different viruses are very similar. Diagnosis is made by examining antibody titres in paired serum samples at 10-day intervals. The virus may be isolated from nasal swabs. Symptons include a short fever (39 to 41°C/102 to 106°F) lasting one to three days, depression and lack of appetite, a nasal discharge and a cough.

The virus invades the lining cells of the windpipe, nose and lungs, causing acute inflammation. Antibiotics are not effective against viruses but can be used as supportive treatment to prevent secondary bacterial infection of the weakened horse. The horse should be confined to its box (which should be clean and airy) and kept warm until its own body defence mechanisms overcome the viral assault.

It must be remembered that the internal damage may still not be healed until 3 weeks after the clinical signs of illness have gone. A rule of thumb is to give the horse a week off work for every day that it has had a fever. If the horse is worked too soon, before the cough has completely subsided, the ulcers in the lining of the nose and throat caused by the virus become infected. This can cause high fever, a pus-like nasal discharge, and even pleurisy if the horse is stressed. If permanent damage is to be avoided, a horse may need at least 3 to 4 weeks after even mild respiratory infections.

Influenza

Equine influenza is a highly contagious respiratory disease caused by a myxovirus, of which there are several types. The 'flu can vary from being mild (almost unnoticeable) to severe, depending on the age and condition of the affected horse and the type of virus. After a short incubation period, the horse develops a temperature lasting 1 to 3 days and a characteristic cough. this starts as a dry, hacking cough soon after the onset of fever, then becomes moist and less frequent, persisting for several weeks. The horse may also have a watery nasal discharge and show weakness, stiffness, loss of appetite and depression lasting 2 to 7 days.

Secondary bacterial infection is rare but can occur, leading to pneumonia, chronic bronchitis, emphysema and myocarditis. 'Flu can be positively diagnosed by the increasing level of antibodies in the blood between the acute and the convalescent stage. Vaccines are available, and they have considerably reduced the incidence and severity of 'flu attacks. Vaccination is advisable for all horses that attend events such as hunting, where strange animals are brought together, and is compulsory for many equestrian sports including racing, horse trials and affiliated showjumping; however, the rules on vaccination certification vary between racing and other sports.

Equine Herpes Virus 1 (Rhinopneumonitis)

This virus causes acute respiratory disease symptoms similar to those of equine influenza. Secondary bacterial infection may lead to a pus-like nasal discharge in more serious cases, but the horse usually recovers in 10 to 21 days.

Herpes myelitis may result in posterior inco-ordination or even hind limb paralysis. This virus may also result in the 'loss of performance syndrome', which may last at least 10 days.

Adenovirus

This virus affects the upper respiratory tract and conjunctiva, causing a mucus discharge from the nose and eyes accompanied by coughing, fever, etc. This can be fatal in foals.

Equine herpes virus 1 and adenovirus may occur in a mild form, called 'snotty nose' when the fever goes unnoticed and only the nasal discharge is seen. This is common in foals and yearlings and may continue for months, but is rare in adults.

Epistaxis (Bleeders)

Horses that suffer from epistaxis bleed from the nostrils. The source of the bleeding may be the mucous membrane of the nasal passage, throat, gutteral pouch or lungs, and bleeding may occur for several reasons:

(1) Gutteral pouch infection, usually fungal.
(2) If for any reason the horse holds its breath or breathes irregularly, air will not move smoothly and may remain trapped in the lungs.

This causes the alveoli to inflate, slowing the flow of blood out of the lungs. Back pressure develops in the veins, and the walls of the blood vessels become damaged and bleed into the lungs. However, the theories on the aetiology of idiopathic pulmonary haemorrhage are still wide open.

(3) Inadequate recovery time allowed after 'flu attack. The lesions in the respiratory tract bleed when the horse is galloped.

(4) Ethmoid haematoma.

There is no reliable cure for epistaxis, although coagulant drugs, vitamin C and antibiotics may help. Rest is always necessary to allow time for lesions to heal.

Loss of Performance

There are many other viruses which can affect horses, and although they may not have such dramatic and recognisable symptoms they are the cause of horses not performing as well as expected (loss of performance). Many racehorse trainers have their horses regularly blood-tested and will not race an animal 'if its blood is not right'. However, it is a widely held misconception that blood-testing helps identify that a horse has a viral infection before the clinical symptoms appears. The parameters examined in the blood test are white cell total and differential counts. These are a reflection of the horse's response to the infection, and thus there is always some time lapse.

12 Heat Stroke and Dehydration

Great Britain may be considered as having about the best climate in the world as far as competing horses are concerned – something to be borne in mind during the occasional long winter and inclement spring. However, recently horses have been competing in hot and humid conditions, conditions bad enough to cause concern for their welfare. This has been highlighted by the problems encountered at Lexington (1984), Bramham (1984) and the Olympics in 1984. It would seem likely that, as the eventing season gets longer and the summer break is eaten away, horses will be competing more frequently during the summer when temperatures are highest. It must be remembered that horses working and travelling at speed in hot and humid conditions – for example, showjumpers and endurance horses – are equally at risk.

Heat Loss

During muscular exertion a great deal of heat is generated in muscles. At a heart rate of 140 to 160 per minute, a 1°C temperature rise every 3 minutes would be generated in the muscle if there were no heat regulation in the body. Normally, however, the horse's body temperature is maintained within a very narrow range by sensors situated in the body which stimulate protective reflexes to raise or lower temperature. If extreme conditions cause these controls to become inadequate and the body temperature becomes too high or too low, the horse becomes ill. If the conditions persist, the horse may die.

Methods of Heat Loss

There are three main ways in which the horse can lose heat:

(1) *The skin.* The circulatory system takes heat from the muscle to the surface; the superficial blood vessels dilate so that more blood can circulate close to the skin and heat is lost to the environment (fig. 12.1).

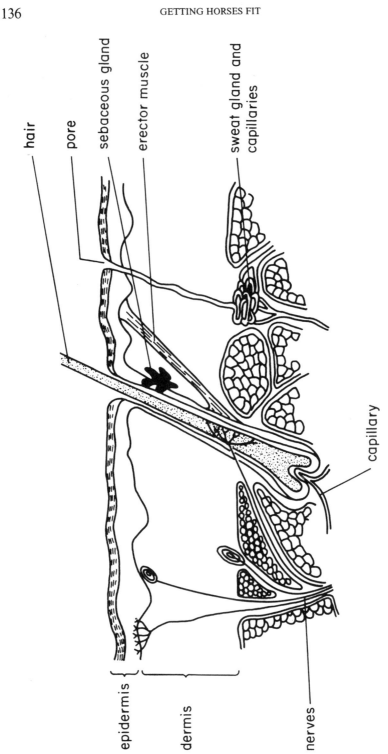

Fig. 12.1 Section through the skin

(2) *The lungs*. The air breathed out is relatively warm and has a high water content – higher than that breathed in; it is a useful form of heat loss and in those animals that are unable to sweat it is the major method of losing heat, e.g. dogs panting. Horses only pant if conditions are extreme and cooling from the skin is ineffective. They may pant up to 200 times per minute to increase heat loss.

(3) *Sweating*. Sweating involves the evaporation of liquid, excreted from the sweat glands, from the surface of the animal (fig. 12.1). Heat is used up when moisture evaporates into the air, so that the surface from which it has evaporated – in this case the horse's skin – is always left cooler. Sweating is a very efficient cooling mechanism.

All these processes take place more efficiently when the air temperature and humidity are low and the air moves, i.e. there is a breeze. Any insulating materials, such as rugs, tack, a thick coat and fat, make heat loss difficult. If the horse's coat is wet, i.e. there is visible sweating, then effective heat loss is not taking place because the moisture is not evaporating quickly enough. Humid conditions, where the air already has a high water content, make effective sweating difficult and can be a serious threat to the competing horse. When looking at weather conditions, it is the *effective temperature* that is important, i.e. the combination of ambient temperature and relative humidity (how much water the air contains). Table 12.1 shows heat loss ability (effective cooling) at different effective temperatures.

Table 12.1 Heat loss ability at different effective temperatures

Ambient temperature plus relative humidity	Effective cooling
Less than 54°C (130°F)	No problem
More than 60°C (140°F)	Increased sweating
More than 66°C (150°F)	Effective sweating lowered
More than 82°C (180°F)	Cooling from skin ineffective, horse pants

Dehydration Through Sweating

Horse sweat is very rich in electrolytes, which are mineral salts contained in blood plasma (the liquid part of the blood from which sweat is derived). Electrolytes must be present in the body in the correct proportions so that normal metabolism can continue. Horse sweat has an electrolyte content ten times that of human sweat (table 12.2) and is also more concentrated than blood plasma in these salts. This means that while in the human marathon runner water replacement is most important, in the horse it is also very important to ensure replacement of electrolytes.

Table 12.2 The composition of sweat

	Sodium(Na) grams/litre	Potassium(K) grams/litre	Chloride(Cl) grams/litre
Plasma	140	3.5–4.5	100
Human sweat	10–60	4–5	10–60
Horse sweat	130–190	20–50	160–190

Horse sweat is also very high in protein – 15 to 20 grams per litre, compared with blood plasma which contains 60 grams per litre. However, this protein is only lost during the early stages of sweating. It is stored in the sweat glands and has detergent-like properties, i.e. it allows the sweat to spread along the hair, causing more effective evaporation. This protein causes the white lather seen on profusely sweating horses, e.g. racehorses parading in the ring.

The amounts of water and electrolytes in the body are controlled by the kidney altering the amount and composition of the urine. Losses of water and electrolytes through sweating and panting are beyond the control of the kidney, and must be replaced through eating and drinking. A horse can lose 15 litres (about 3 gallons) of sweat in an hour of hard work in hot conditions, and may become dehydrated if this is not replaced. The horse has a great problem in maintaining normal physiological conditions inside the body when working in hot, humid weather.

The Effects of Fluid and Electrolyte Loss

By the time a horse has sweated 22 litres (5 gallons) of moisture which it has not replaced by drinking, it will be suffering from dehydration. The fluid is lost from blood plasma (extracellular fluid) and from inside the cells (intracellular fluid). This upsets cell function, which means that tissue and organ function will also be impaired. The blood supply will have diminished, and the blood will be more concentrated and thicker.

This has four effects:

(1) As blood volume falls, there is less sweating and an increase in body temperature.
(2) The oxygen-carrying capacity of the blood is diminished. The muscles will become increasingly short of oxygen, and in order to keep functioning an alternative source of energy supply is required. This is obtained by breaking down energy stores in the muscle without the use of oxygen (anaerobic respiration). Lactic acid is released, resulting in failure and possibly destruction of muscle fibres. This is manifested as azoturia, set-fast or tying-up. The horse may also exhibit colicky pains.
(3) As cooling from the skin becomes ineffective, the horse will begin to pant. If the respiration rate is greater than the pulse rate, the horse is suffering from heat stress, i.e. there is too much heat in the body, and veterinary help should be sought.
(4) Sweating causes loss of electrolytes, and this is responsible for some of the signs of dehydration and exhaustion including quivering of the muscles and 'thumps' (synchronous diaphragmatic flutter), when a horse's heart appears to be beating in its flanks.

Temperature

A horse's normal temperature is 38°C (100.5°F). During exertion the horse has the ability to store some heat, i.e. the core temperature in the centre of the body rises. It is not uncommon to see temperatures of 39°C (102°F) at the end of Phase C and 40°C (103°F) at the end of Phase D of a three-day event. There is no cause for concern unless the temperature is greater than 40°C (103°F) in very hot watear, or the horse's temperature is static or rising. A bright, alert horse would be allowed to start phase D at this level. A horse with a temperature of 41°C (104°F) plus should on no account be allowed to continue. Remember that these are all rectal temperatures. The core temperature will be about 2° higher. Muscle fibres start to die at about 45°C (108°F), and any horse with a temperature higher than this will suffer muscle damage – if indeed it survives.

Assessment of Dehydration

Dehydration can be assessed by:

- the pinch test
- biochemically

If a fold of skin is picked up, and does not return in a normal supple manner but remains standing up the horse is dehydrated. This is easiest to test on the shoulder or neck.

Dehydration can be detected by measuring protein in the plasma. Obviously, a blood test takes time and this is not always a satisfactory method.

Reducing the Occurence of Heatstroke, Exhaustion and Dehydration

It is essential that every rider is aware of the symptoms of dehydration and exhaustion. Some of the very severe signs have already been mentioned, e.g. panting, quivering of the muscles and 'thumps'. In addition, all tiredness is accompanied by general listlessness and a lack of inclination to eat or graze. A horse whose muscles are seriously short of oxygen may breathe so slowly that he takes less than one breath per stride; a horse that is also overheated may take several short breaths between long breaths until oxygen needs are met, and then panting may be almost continual.

Dehydration is indicated by patchy thick sweat, which is very serious because the animal will then be at risk from overheating. There will also be a loss of the normal colour from the membranes of the eyes and gums. If a finger is pressed to the gum, the blanched area made takes a longer time than normal to re-colour. This indicates that the blood is too thick to circulate through the small blood vessels and that rate of flow has slowed down. If a fold of skin on the neck pinched between finger and thumb takes more than 5 seconds to flatten out, this indicates that the horse is moderately to severely dehydrated – riders can use this test as they go along.

Should any of these signs become apparent, the rider should slow the pace and seek veterinary advice as quickly as possible as to whether or not to continue.

The best way to avoid these problems can be considered by looking at:

• the horse
• its preparation
• its management during the competition.

The horse should be the right type for the job required of it. Many long-distance horses are Arabs or of Arab breeding, and the most successful horses have only been about 15 hands high. In Britain we had the perfect long-distance horse in the Norfolk, but recently allowed it to die out. The best three-day event horses tend to be Thoroughbred or $\frac{7}{8}$ths Thoroughbred, lean athletic types with flowing, economical action.

The object of preparation is to produce an animal athletically capable of the task, mentally content and physiologically in a normal state. However, the horse may be subject to dehydration even during training. The onset of a shortage of water can be insidious – for example, it may be brought about by standing in a hot lorry for a few hours. Fit horses should have water made available to them every hour, particularly if they have worked first. Care should be taken that the horse drinks the water offered to it during its stay at a competition. This may mean bringing water from home, or adding common salt or beet molasses to the feed, which may help the reluctant horse to drink.

The management of the horse immediately before and during the competition is of great importance. The eventer on cross-country day should have water freely available until an hour or so before Phase A. Horses' stomachs empty very quickly and it is not necessary to withhold water for very long before starting. If a horse has free access to fresh water in its box, it is unlikely to drink too much at once.

If possible, avoid hurrying your horse into the box at the end of Phase C, as it will have much more recovering to do and the halt loses much of its value. While remembering to give the inspection panel 25 metres trot as the box is entered, it is best to walk the previous hundred metres or so. When in the box, the horse should not be allowed to stand and stiffen up.

Under normal conditions, the neck and extremities should be sponged with water which is allowed to evaporate. Remember that it is not the application of water that cools the horse but its evaporation from the skin, thus warm water is used to keep the surface blood vessels open. If cold water is used, it may cause the blood vessels of the skin to constrict; heat is not lost so effectively and builds up. As soon as the horse is dry, water should be re-applied.

Under hot conditions, water should be splashed over all the body except the croup and loins, where the result may be stiffening and even tying-up of the back muscle. Ice should also be available under severe conditions and applied to the head, throat and inside legs, where it can cool the blood running through surface vessels. A mixture of spirt and water cools and cleans very effectively. A dry numnah should be fitted to the saddle if the other numnah is soaked through with sweat. A small mouthful of water may be offered – more will merely be an unwanted burden in the stomach during Phase D as water will not be absorbed from the stomach to a significant degree during the cross-country phase – or the horse's mouth may be sponged out (fig. 12.2).

After the application of these precautions, your horse has a better chance of being ready to gallop over a demanding course with a smaller risk of suffering from the heat.

Fig. 12.2 Sponging out a horse's mouth

After the cross-country, the horse should be treated in much the same way as during the 10-minute halt. The importance of walking the horse cannot be too strongly stressed – this keeps the blood flowing through the body, helping disperse metabolic waste products formed during galloping, and also helps the animal cool. The horse should be sponged down, walked until it is dry, sponged again and walked slowly until the pulse and respiration rates are approaching normal. Water may be safely offered at a rate of 4 litres (1 gallon) every 15 minutes (this is discussed in a later chapter). However, it must be remembered that a tired and possibly dehydrated horse may take a long time (up to a minute) to decide to drink, so always wait when offering the horse water and do not take it away too quickly.

If the climate is very hot and the horse is not accustomed to the degree of heat, e.g. horses from the U.K. competing in Kentucky or California, dehydration may set in before the competition has started. It may be necessary to cool the stable by waterspray on the roof and fans at the door to create a breeze. Electrolytes should also be used, and these will be discussed later.

Keeping cool and the provision of water are also very important for the endurance horse. Before the ride, water must be freely available in the box and paddock. There will then be no need for the horse to drink excessively before a feed or before work, nor will it start work thirsty. During competition, the horse should ideally never become excessively thirsty. However, this is not always possible due to the speed, the route or the weather, and it may become impossible to keep the horse so well watered that it arrives at each drinking place in a fit state to drink all it needs without pause. In these cases it is essential that the horse should remain by the water, drinking at intervals until the thirst is quenched. If the thirst is not quenched, the horse will be even more thirsty by the time the next drinking place is reached.

It is advised that drinks of 15 to 20 swallows (about 2 litres or 4 pints) followed by a pause of 1 minute should be allowed. The horse should then be ridden at walk for several hundred metres until the water in the stomach has had time to warm up. Provided these methods are followed, there is no danger in giving a hot horse cold water.

At a rest halt, the horse should be washed down as previously described, taking great care to clean properly under the saddle and girth. The legs and saddle area may be rubbed with methylated spirit, which restores circulation and quickly evaporates, leaving the surface fresh and cool. Nostrils, eyes and dock should also be well sponged with clean water.

If the horse has been allowed to drink properly during the ride, it should not arrive at the halt excessively thirsty and should be allowed 2 to 3 litres (4 to 6 pints) every 5 minutes until the thirst is quenched. If the horse is kept moving there is no danger in giving cold water, but most riders prefer to give water with the chill taken off it (known as 'chilled water').

Electrolytes are powders or liquids given in solution in the drinking water before, during and after competition to replace electrolytes lost during sweating. A horse's diet may, as previously mentioned, be low in salt, and if the horse sweats it may suffer a sodium chloride deficiency. This may be partially overcome by providing a self-feed salt lick – this is not sufficient for horses in very hard work, but it can be regarded as insurance to top up a horse's salt levels. After a competition, it may take 5 to 6 days to replace the potassium lost during heavy sweating. A potassium deficiency is known to be a factor in the human equivalent of azoturia and may also have an effect on horses.

There is a school of thought which believes that giving electrolytes 'trains' the body not to make up its deficiency from the diet and maintains that correct food is all that is necessary. Probably the best plan is to use electrolytes for a few days in anticipation, during and for a few days after major stress in warm, moist conditions.

13 Internal Parasites

It is important for horse-owners to appreciate that no horse can be expected to perform at maximum levels when suffering from a worm burden or the damage caused by previous parasite infestations. A regular worming programme from six weeks old continuing throughout the horse's life is essential. The horse is host to a wide variety of internal parasites, only the most important of which will be discussed here – namely large and small roundworms (strongyles), ascarids, bots and lungworm, as these are the ones which tend to affect performance.

Strongyles

There are two groups of strongyles: the large strongyles and the small strongyles. The redworm (*Strongylus vulgaris*), one of the large strongyles, causes the most severe damage of all the internal parasites. The larval stages of the worm travel throughout the horse's body. First, they burrow through the intestine wall and enter the small arterioles of the blood circulation, then migrate through the arteries, travelling against the flow of the blood.

Larvae may enter the aorta and travel to the heart, causing valve damage, or enter the renal artery (supplying the kidney), but the majority tend to stay in the anterior mesenteric artery which supplies the intestines. Here, the larvae attach to the arterial walls, casing inflammation and a weakening and bulging of the wall called an aneurysm. If there is a lot of irritation, the artery may become blocked due to the build-up of larvae and blood clots; this results in part of the intestine being starved of its blood supply and dying (infarction). Food is unable to pass normally through the dead portion of the gut, resulting in colic. Redworm larvae are thought to be responsible for the majority of cases of spasmodic colic in horses.

The life-cycle of the redworm is very long, ingestion of larvae to the appearance of adults in the large intestine taking 6 to 12 months (Fig. 13.1). In the horse's intestine, the female worm lays eggs which pass out

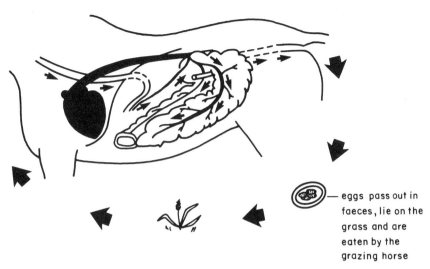

Fig. 13.1 The life-cycle of the redworm (*Strongylus vulgaris*). The ingested larvae enter the arteries suplying the gut and migrate up the vessels to the cranial mesenteric artery. After a period in that vessel they migrate down the arteries to enter the large intestine and become egg-laying adults

with the dung on to pasture; the eggs hatch to give stage 1 larvae. These feed and become stage 2 larvae which in turn become stage 3 *infective* larvae – these, when swallowed by the horse, can complete their life-cycle. The larvae then migrate through the horse's body as described, returning to the large intestine as egg-laying adults. The adult worm can cause damage by sucking blood from the intestine and causing anaemia, ulceration, bleeding, colic or diarrhoea.

Small strongyles are not as destructive because their larvae do not travel outside the wall of the intestine. The adult worm does not usually suck blood and the major damage is caused by the ulcers left in the intestine wall after the larvae hatch from it. Small strongyles mature in 6 to 12 weeks.

Ascarids (Parascaris equorum)

Known as the whiteworm, this is the largest roundworm found in horses, the male reaching up to 30 cm (12 in) in length and being as thick as a pencil. The adult worm lives in the small intestine, where the female lays eggs which pass out in the droppings. Infective larvae develop; they are swallowed by the grazing horse and pass into the intestines. The larvae burrow through the intestinal wall and are carried to the heart, lungs and liver in the bloodstream. After a period of growth and development,

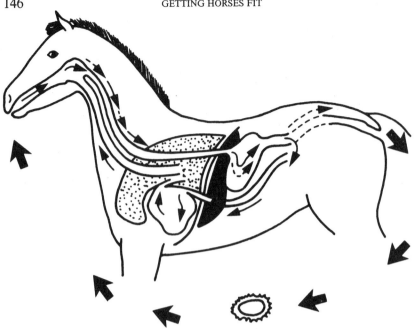

Fig. 13.2 The life-cycle of the ascarid. The eggs are picked up with grass, swallowed, and pass down the oesophagus to the stomach and small intestine. They migrate from the intestine through the liver and lung, are coughed up into the throat and reswallowed. Once again in the small intestine they develop into egg-laying adults

causing damage in these organs, larvae invade the windpipe and are coughed up and swallowed. Larvae pass through the stomach to the intestines where they grow into adults, completing the cycle in about 12 weeks (fig. 13.2). A female can contain 27 million eggs and lay 200,000 eggs a day. The eggs can live for months and are resistant to many chemicals.

Ascarid infection is most common in foals and young horses that have not yet acquired resistance to them. Due to the persistence of the eggs, this year's foals can pass out eggs that will infect next year's foals. Ideally, young foals should be turned out on 'clean' pasture, i.e. grass that has not been grazed by horses during the previous 12 months.

Ascarids cause unthriftiness, lung damage and occasionally rupture of the gut due to the massive buildup of worms. This is fatal.

Ascarid infections are fairly easy to control; foals should be wormed at six weeks old and then at 4 to 6 week intervals.

Bots

These are the larvae of the botfly (*Gasterophilus*) which live in the stomach of the horse. The fly is about $\frac{1}{2}$in long and bee-like although it cannot suck, bite or feed because its mouthparts are degenerate. The fly is active in early summer and September, making a humming sound as it flies, and lays its yellow eggs on the hair of the throat, legs or on the lips. Each egg hatches in 9 to 12 days but the first larvae cannot develop until they are licked into the warm, moist horse's mouth. In the mouth the larvae burrow into the cheek or tongue for 20 to 30 days before migrating to the stomach. The second and third stage larvae (*bot maggots*) grow in the stomach living on the host's food for about 12 months. They then pass out with the droppings, the maggots pupate for about three weeks and the adult fly emerges, thus completing the cycle.

The adult fly can cause horses to panic and gallop about. The maggots can cause inflamed ulcers in the stomach and thus disrupt digestion.

Lungworm (Dictyocaulus arnfieldi)

These are roundworms living in the air passages of the lungs where they lay their eggs. These are coughed up and swallowed and pass out in the droppings where they hatch and are picked up by another horse. Once swallowed, the larvae bore through the intestinal wall and enter the lymph stream. Venous blood carries them through the heart to the lungs, where they develop into egg-laying adults. The worm irritates and inflames the bronchi of the lungs causing bronchitis and a chronic cough. Migrating larvae can cause scouring. Donkeys are capable of carrying large numbers of lungworm without showing any symptoms and should not be allowed to graze with horses unless they have been checked to see if they are carriers, and then treated.

Diagnosis of Worm Infestation

Although a horse may be regularly wormed, it should still be checked at intervals for the presence of a worm burden because the parasites may have built up resistance to the chemical being used. This check can be done in two ways: a worm egg count in the dung or a blood test.

Egg Counts

A sample of fresh dung is looked at under the microscope and the number of worm eggs counted. The presence of even small numbers of eggs in the dung indicates large numbers of egg-laying adults in the intestines. This test can, however, be misleading – if a horse has been recently wormed there will be no eggs in the droppings, but there may still be many larvae migrating through the tissues, providing a back up ready to enter the intestines and mature. The horse is not free of these parasites: the dangerous larvae are still there.

Blood Tests

The presence of migrating larvae in the tissues will cause changes in the white blood cell numbers. In particular, eosinophil numbers increase. Eosinophils multiply in allergies; an allergy is an intolerance to foreign protein, so they increase when there is a worm burden. Eosinophil numbers will not return to normal until the migrating larval stages have been removed. However, eosinophils will increase in other allergic reactions and it may be necessary to look at the serum protein levels. Normally the albumin:globulin ratio is about 1.0:0.7, 3 g/100 ml albumin and 2.1 g/100 ml globulin. When there is worm infestation, the globulin levels increase so that the ratio may be reversed, eg. 0.5:1.0.

Treatment of Worm Infestation

There are many commercial wormers available. The horse-owner must read all labels on commercial products to determine whether they are effective against the adult stages of strongyles and ascarids and include active ingredients such as:

> *Thiabendazole* (Equizole) safe for pregnant mares
> *Mebendazole* (Telmin)
> *Pyrantel tartrate* (Strongid)
> *Dichlorvos* (Equiguard) effective against bots
> *Phenothiazine* (Cooper's liquid) not effective against ascarids
> *Piperazine* (Coopane) not effective against redworm

(This list is not complete, and is only given as an example.)
Safety when fed to foals and pregnant mares should be checked if necessary. Intervals between treatments must be carefully gauged, because larval forms are not harmed by these anthelmintics (wormers).

The most effective worm control is obtained by worming every 4 weeks. This is essential on grass heavily grazed by horses, but is only really worthwhile if all the animals are wormed. Frequently, a 6 to 8 week worming regime is adopted. This is the minimum time taken for small strongyles to mature and is satisfactory on good grazing which is not horse-sick. Worming spring and autumn is *not* adequate.

It must be remembered that it is equally important to worm the stables horse at 6 to 8 week intervals. During the horse's spell at grass it will have picked up redworm larvae which may take 12 months to become egg-laying adults. A horse that has been fully stabled for a year may therefore still be carrying a worm burden. This emphasises the long-term effect that an inadequate worming programme at grass can have on the stabled horse – all the time redworm larvae are migrating through the horse's body, they are causing damage.

Recently, there has been the development of a revolutionary anthelmintic called ivermectin (marketed as Eqvalan). This drug kills internal and external parasites including the migrating larval stages of redworm and bots. Regular use can lead to a much more effective worming programme of the stabled and grass-kept horse.

Part IV

The High-Performance Horse

14 Care of the Performance Horse

Stress

The word 'stress' occurs frequently in talking about competition horses, and it is important to understand the implications of stress.

Any competing horse is under stress, and the best performers may be those best able to tolerate it. The rider and trainer must be concerned about the degree of stress a horse can be exposed to without causing harm and suffering. It must be remembered that stress is a natural and normal respose to daily variations in the environment. Within limits, the stress response will help, ensuring *adaptive responses* – this means that the next time the same situation arises the horse will not find it so stressful because it has previously adapted to that situation.

The most obvious example of stress a horse is exposed to during training is a progressively increasing workload. Each time the horse is worked, a new level of stress is introduced, leading to fitness to cope with a predetermined level of exercise. In other words, stress is a necessary and desirable part of a horse's training programme. If the horse cannot cope with the level of stress it is being exposed to, it becomes *distressed* – obviously, overstress is not desirable.

Many performance horses are subject to repetitive competition stress, and the rider and trainer must remember that each competition involves more than the physical stress of jumping or galloping. The stress of travelling and the mental stress of being in a crowd and away from home will also take their toll.

Competition also involves the grouping together of large populations of horses, any of which may be suffering from obvious illness or subclinical disease (the horse is ill but showing no symptoms). Chronic stress reduces a horse's resistance to disease, a factor which may explain why so many horses succumb to viral infections. This is particulary true in racehorses, where the animal is under the added stress of growing and performing at the same time.

There is no doubt that the stressful conditions imposed on competition horses are most often caused by lack of knowledge of ability on the part of the rider or trainer. This leads to poor management, training and riding, exacerbating the stress the horse is already under. Riders and trainers must be aware of an individual horse's ability to cope with stress and treat each horse accordingly. This is where the 'art' of training takes over from science – anybody can get horses fit, but not everybody can keep them that way.

Care After the Event

After the horse has finished a high-stress competition, it will benefit from efficient and thoughtful treatment. The horse's temperature, pulse and respiration will all be elevated and it is important to get these systems back to normal as quickly as possible.

Keep the Horse Moving

Immediately after the work has stopped, the rider should dismount and loosen the girth. Now is not the time to untack – it is more important to keep the horse walking so that the circulating blood will cool the muscles and carry toxins such as lactic acid away from the muscles. Standing reduces circulation so that lactic acid may be 'trapped' in the muscles, causing stiffness and perhaps inducing azoturia (tying-up) or even colic. After 5 minutes walking the horse can be untacked, a sweat rug or cooler put on and walking continued. If the saddle has been on for a very long time, it should be left on for longer than 5 minutes to allow the circulation in the blood vessels under the saddle to return. Sudden removal of the saddle can cause pressure lumps and scalded backs.

Check T.P.R.

The temperature, pulse and respiration should be taken to see how much above the normal they are and to monitor how quickly they drop. These values are an indication of how severely stressed the horse is, and how quickly it is recovering. The pulse can be taken under the jaw or using a stethoscope for 10 seconds – multiplying by six to get the pulse rate per minute. The temperature may be up to 43°C (106°F). The quality of respiration should be noted; a horse breathing deeply is taking in oxygen, while rapid, shallow breaths indicate that the horse is trying to cool down.

On average, in a fit horse, it takes about 15 minutes for the pulse and respiration to return to comfortable limits. the horse should be checked every 15 minutes until the values return to normal, which should be within an hour of completing exercise.

Washing Down

The horse should be walked for 15 minutes until the breathing has slowed, and then stood in a sheltered place – out of the wind on a cool day and out of the sun on a hot day. If the horse's temperature is no more than 41°C (104°F), tepid water will cool down the horse quite safely. Ice water might cause the blood vessels near the skin to contract so that the heat is unable to escape from the muscles to the surface. The lower neck, inside the legs, the head and the belly should be sponged. The large muscle of the back and quarters should be avoided, as sudden cooling may send these muscles into spasm.

If the horse's temperature is between 41 and 43°C (104 and 106°F) and the weather is hot and humid, the horse must be cooled as rapidly as possible. Ice water can be used on the horse in strategic places, e.g. under the tail, between the hind legs and on the head, where major arteries pass close to the skin, thus having a maximum cooling effect. The loins and quarters must be avoided. It is important to keep the horse moving, one person leading and one sponging the horse. If the muscles are shaking from stress, the horse must be walked as it is sponged.

Water Provision

The horse must not be given water until the temperature has dropped to 40°C (103°F) and the pulse and respiration have slowed. When a horse is exercising, adrenalin stimulates the blood vessels to the stomach and gut to contract, sending blood to the muscles. While the horse is cooling down, the blood must be kept in the muscles. If it drinks cold water the blood vessels to the stomach dilate and blood rushes to the stomach, away from the muscles, legs and feet. The horse will cool less efficiently and may even become colicky. Once recovery is under way, and until the horse is completely cool, five swallows of water for every 50 yards walked is adequate.

Checking for Injury

During untacking and sponging, the horse must be checked over for scratches and minor injuries which may be swollen and infected by the morning. Any injuries found should be thoroughly cleaned and dressed with antibiotic cream or powder, but bandages must not be put on until the horse is cool. While the horse is being walked around it should be jogged for a few yards to check for soundness.

Once cool and dry, the horse can be rugged up and returned to its box

Fig. 14.1 A stable bandage

to rest. The legs can now be poulticed or bandanged (fig. 14.1). Some people like to use a liniment and give the horse's legs a gentle rub-down to help circulation and ease tension, while others use cooling treatments such as clay poultices.

Keep Checking

The horse may break out in a sweat because although the skin and surface muscles are cool there may be heat and toxins deeper inside the muscles which the body needs to get rid of. The horse should be looked at every 10 minutes for the next hour, checking for cold sweaty patches, restless behaviour, disturbed bedding and reluctance to eat or drink. If all the lactic acid has not been removed from the muscles, these signs could indicate the onset of azoturia or colic. If the signs of discomfort are only mild, the horse should be kept warm and walked in hand until the sweat has dried and the horse comfortable. If the horse is pawing the ground and looks very distressed, veterinary advice should be sought.

The horse must be checked again last thing, and if at all worried it is wise for the rider or groom to look at the horse again during the night. It is sometimes advisable to walk the horse for 5 minutes at the check to ease any stiffness it may be feeling.

Feeding

Many riders worry about the amount their horses eat during a two or three-day event competition. However, there is no need to worry too much. The horse has more than enough stored energy in its liver and muscles to see it through the competition next day. Do not depart far from the horse's normal feed but, add high-energy, palatable feeds including flaked maize, dried milk and boiled linseed.

The fluid and electrolyte balance is very important, and the horse must be constantly watched for signs of dehydration. If a pinch of skin on the neck lingers after it has been released, and the horse has a gaunt, tucked-up appearance, it may well be dehydrated – this really can limit the next day's performance. Ensure that the horse drinks, and provide electrolytes in the water and feed.

Colic can be a problem after severe exertion, and the intestines must be kept moving. Once the horse is cool it may appreciate a small bran mash, with its normal ration given later on. The tired horse is easily overfaced by a large feed, but dividing the normal feed into two and feeding it at intervals may overcome this.

Signs of Overstress

(1) *Heat, swelling and pain* in a specific area, indicating injury.
(2) Reluctance to eat, restlessness, listlessness – any sort of *abnormal behaviour* may indicate the onset of colic.
(3) *Shaking* indicates acute muscle fatigue, and is the result of toxic lactic acid build-up and electrolyte imbalance in the muscle. The horse must be kept warm and walked in-hand to keep the blood flowing through the muscle – otherwise the horse may tie-up.
(4) If the horse is *reluctant to move* or get up and appears to be in pain (increased pulse and respiration), it may be suffering from azoturia or laminitis. The vet should be called if the signs persist.
(5) *Thumps* (synchronous diaphragmatic flutter) is a condition where the horse shows rapid breathing with a distinctive thumping in the flanks as if the heart were beating there. Thumps occurs in horses that are nearly exhausted and is caused by an electrolyte and acid

imbalance in the body. Immediate veterinary attention is necessary.

(6) The horse's *appetite* will tell you how tired he has been. Until the horse is eating up normally, it has not recovered from its exertions and should be allowed plenty of rest. Hacks and grazing in-hand will help the horse relax and recover.

Measures to Take to Reduce Stress

Always *warm up* the horse properly. The muscles should be supple and the horse alert and moving freely. Warming-up for too long exerts the horse unnecessarily, using up valuable energy stores, while lack of warming-up may mean the horse has to compete with the body systems unprepared, increasing chances of injury.

Assess the horse. It is vital to know how fit your horse is, and at what speeds within paces it works best; this way you can work the horse to give its best performance without overstressing it. Judge the pace so that the horse is not too tired, and adjust the speed if necessary. It is the rider's job as well as the vet's to ensure that the horse is fit to continue.

Use the rests and breaks between classes or phases of the competition intelligently. The procedure at rest stops should follow the same lines as the early stages of cooling the horse down after the competition. The horse must be kept moving – a 20 yard walk every minute or so would be adequate. A rug should be thrown over the horse if the weather is cool, and shade found if the day is hot. Temperature, pulse and respiration should be checked. The horse can be allowed a few swallows of water if it is not badly overheated; most horses in endurance competitions learn to drink when water is offered. Tack should always be checked for safety and comfort, and changed if necessary.

Take the horse's *normal feeds* with you to the competition. Stick to the same brands so that there is no sudden change in the horse's diet which could upset its appetite or digestive system. Although bran is currently out of favour with nutritionists, an occasional bran mash is useful before rest days or after hard work.

The *routine* at competitions should be as close as possible to the routine at home. Horses are creatures of habit, and thrive on routine.

The horse should be allowed to *relax* as much as possible at a competition. Grazing in-hand can accomplish this.

The horse must be thoroughly checked *the day after* for soundness – the legs looked at, the horse jogged and hand-walked or turned out to roll and graze. The temperature should be taken; a rise in temperature could indicate illness. Horses weakened by the stress of exertion are more susceptible to illness and disease.

The horse must be *travelled* carefully. The legs and head should be adequately protected and the vehicle driven with great care. Travelling can be exhausting for a tired horse and long distance journeys should be undertaken thoughtfully, stopping frequently to water the horse and perhaps allowing it to stretch its legs.

In the U.S.A., horses are subjected to very long journeys. Travelling in high temperatures can lead to dehydration which in turn can cause impaction of the intestine and colic. Some vets recommend that the horse has a bran mash the night before and only hay the morning of the journey.

Following these procedures before, during and after competition will help minimise the physical and mental stress that the horse is exposed to. In turn, this will help the horse enjoy a longer and more successful working life.

15   Interval Training

Interval training was originally developed to train middle and long-distance runners and swimmers, and has been successfully used in human athletics for over twenty years. Jack Le Goff (the United States event team trainer) adapted interval training for horses and supervised its use for three-day event horses in America. This, and the use of interval training by top British eventers including Lucinda Green and Virginia Leng, has led to a great deal of interest in this method of training, amongst riders at all levels of the horse trials world.

Interval training, like any form of conditioning programme, needs careful planning and implementation. It cannot cut corners as compared to more traditional training methods, nor can it allow a complete novice to train a horse to three-day event fitness from a book. However, if interval training is combined with the use of measurable indicators of fitness such as pulse, respiration and recovery rate, it does help the less experienced rider get a horse fit, rather than having to rely on that indefinable quality, 'feel', which only comes with years of practice.

What is Interval Training?

Interval training consists of giving a horse a period of specific work, followed by a brief interval of semi-rest during which the horse is only allowed to partially recover from its exertions before being asked to work again.

The aim of Interval Training

As has been explained previously, a fit horse is one which has a greater capacity for using oxygen (aerobic work). In other words, the point when anaerobic work starts (the anaerobic threshold) is delayed. The periods of work given during interval training develop and extend the horse's capacity for intensive work as the programme progresses, without there being a build-up of lactic acid. Lactic acid is a major factor contributing to fatigue; the rest periods allow the blood to remove any lactic acid that may be building up in the muscles. Interval training builds increased tolerance to stress, and the horse's reaction to the training programme can be carefully monitored.

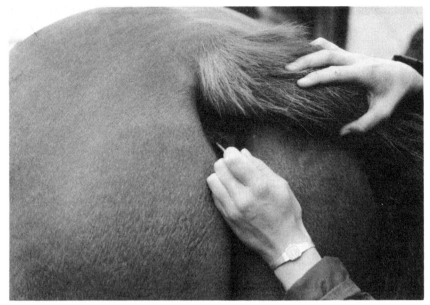

Fig. 15.1 Taking a horse's temperature

Fig. 15.2 Taking the pulse

Fig. 15.3 Using a stethoscope

Monitoring Interval Training

An essential part of the interval training regime is monitoring the horse's temperature, pulse and respiration (T.P.R.). The horse cannot tell us how it feels so it is necessary to learn to read the signs of stress.

Temperature

The horse's normal temperature is 38°C (100 to 101°F). Any deviation from normal may indicate stress of some sort, most commonly illness. A clean rectal thermometer, lightly greased with vaseline or a similar jelly, is used to take a horse's temperature (fig. 15.1). It is placed full-length in the rectum, left for 1 to 2 minutes, removed and read.

Instant digital read-out thermometers are now available and, if they prove reliable, will be most useful.

Pulse

The normal range of pulse rates in the horse is 32 to 44 beats per minute. The easiest place to take the pulse on a horse is on the facial artery which runs around the inside of the jaw (fig. 15.2). A stethoscope can also be used on the left side of the horse just behind the elbow, in front of the girth (fig. 15.3). Count for 15 seconds and multiply by four to get the heart rate per minute.

Respiratory Rate

The normal breathing rate – how often a horse inhales and exhales – is 8 to 16 per minute. To take the respiration rate, the in and out motion of the ribs or rise and fall of the flanks is observed (fig. 15.4). Each combination of in and out is counted as one. The rider can also put a hand close to the horse's nostril and count as the horse breathes out. To obtain horse's normal values, each of these measurements should be taken while the horse is calm and at rest; for example, between feeding and the daily exercise. These normal values can then be used as a base line to compare against the values obtained after exercise.

How to Use Interval Training

Interval training cannot cut corners. The following factors must be

Fig. 15.4 Recording the respiration rate

considered before starting an interval training regime:

(1) There must not be any lameness or any doubts about soundness.
(2) The horse must have been regularly wormed.
(3) The teeth should be regularly rasped.
(4) It is useful to have a blood test taken to ensure the horse is not anaemic.
(5) The horse must be eating well.
(6) The horse must have had six weeks' basic fitness work and should be capable of 90 minutes of walk and trot over rolling terrain without distress. The horse conditioned slowly and carefully will stay in peak condition longer than one pushed too fast in the early stages.
(7) It is essential to keep a notebook with a running record of the horse's response to training so that the programme can be adjusted accordingly. As well as pulse and respiration rates, the weather and type of work and how the horse 'feels' should also be recorded. The weather can have a profound effect on how quickly the horse recovers from work. During hot, humid weather the horse's respiratory rate will take longer to fall after exercise because more rapid breathing is necessary to reduce body heat; so even though the pulse rate may drop to normal, the horse's breathing may still appear high. Pulse is, therefore, a more reliable gauge of condition, and is the most important recording to make.
(8) As in any form of training, the rider must be alert to any change in the horse's attitude, appetite, coat, droppings, appearance, muscle tone, etc.
(9) Interval training may not be suitable for all types of horse. Youngsters, particularly, may not be physically or mentally able to cope with a rigorous training schedule. Some excitable horses seem to settle into the canter workouts, making them easier to hold when competing, while lazy horses get bored with interval training.
(10) The breeding, sex, condition and previous training must also be considered before starting a training programme.

During exercise, pulse and respiration rates increase. The horse is given a set piece of work and the pulse and/or respiration rates recorded immediately after work and 10 minutes later. The difference in the two readings indicates the 'recovery rate' of the horse. This piece of work is repeated twice a week and the pulse and respiration rates recorded at the same points; as the horse gets fitter, it will recover faster from the work. Fitness is gradually built up, slowly increasing the amount of work the horse is asked to do by increasing the speed, distance or time of the workout or using more demanding terrain. If the recovery rate is not good enough after a workout, the work load is adjusted so the horse is never overstressed.

The trainer should know the horse and understand its capabilities and limitations, and should be aware of the degree of fitness that is eventually needed so that a flexible programme can be made to suit any individual animal. This is a markedly different system from the traditional one, where horses are trotted and cantered for long distances with short sharp gallops brought in later in the training programme and no repetition of work that day. That method fails to develop fully the horse's aerobic capacity.

The Method

In order to carry out interval training correctly, a *known distance* must be covered in a *set time*. This distance can be over an all-weather gallop,

Fig. 15.5 Interval training round the edge of a field

round the edge of a field or anywhere suitable for doing canter-work (fig. 15.5). Ideally, 1600 metres (1 mile) should be marked off in 400 metre segments; if this is not possible, one 400 metre (0.25 mile) gallop will be adequate. An easy-to-read wrist stop-watch is needed.

The time allowed for covering this distance can be calculated according to the competition that is being trained for:

> 400 metres in 1 minute 49 seconds is 220 metres per minute (m.p.m.), which is the good ongoing trot required for Phases A and C (roads and tracks) in a three-day event and is approximately 13 kilometres per hour (9 miles per hour).
> 400 metres in 1 minute 4 seconds is 350 m.p.m.
> 400 metres in 1 minute is 400 m.p.m.
> 400 metres in 57 seconds is 425 m.p.m.

It is important for the event rider and the long-distance rider to be able to pace his horse over a given distance and different ground during competition, and interval training helps the rider to do this early on in the training programme. It also means that the horse is used to travelling at the required speed, which is useful during the actual competition. In order to design an effective training programme, the rider needs to know exactly what is required in competition in terms of speed, distance, length and intensity, so that these aspects can be incorporated into the programme.

Interval training at trot can begin to be incorporated in week 4 or 5 of the fitness programme, initially serving to accustom both horse and rider to a 220 m.p.m. trot. An average horse trots at about 200 m.p.m. and may need to trot more briskly to cover 400 metres in 1 minute 49 seconds. In the first session, two 400 metres trots with a 3 minute walk in between would be sufficient. After the second trot, the horse should be dismounted; pulse and respiration should be recorded and then checked again after 10 minutes' walking.

After this break the horse's pulse and respiration should have returned to normal – if they have not, the horse is not fit enough and should stay at this work level for the next workout in 3 or 4 days' time. If, however, the horse responds well, a further 400 metre trot can be added after a 3 minute rest at each twice-weekly workout up to a maximum of five repetitions, so that by week 6 to 8 the horse should be ready to canter.

This work at trot is invaluable for introducing the newcomer to interval training, developing a feel for speed and distance, and practising taking pulse and respiration and keeping the records necessary: pace used, length of rest interval and number of repetitions. It is important to remember that the horse must always have at least 15 minutes' warm-up at walk before interval training at trot.

The horse's pulse and respiration will rarely return to the resting values obtained in the stable due to excitement and the fact that it is recovering at walk, so a value for the horse's warmed-up pulse and respiration should be obtained and used as the baseline value.

After week 6 to 8 (depending how the horse is progressing), interval training is ready to start in earnest, so that by week 10 to 12 the horse should be ready to compete in horse trials. Now that the feel of a 220 metre trot has been well established, the trotting sessions can be governed by time rather than distance and gradually the length of time spent trotting increased.

The horse has been working five 1 minute 49 second sessions. The length of each session should be gradually increased to a maximum of 5 minutes, and initially the number of sessions will have to be decreased. Canter-work will also be introduced now, after a 30 minute warm-up at walk and trot. The measured 400 metre distance should be cantered in 1 minute 4 seconds (350 m.p.m.). This is followed by 3 minutes' walking and another 1 minute 4 second canter, after which the pulse and respiration are recorded. If after 10 minutes' walking the pulse has returned to normal, the horse can proceed to the next level, which is to increase cantering time to 3 minutes. If the pulse has not returned to normal, the horse must stay at the former level until it is fit enough to continue.

The work is gradually built up until the horse is cantering three 3 minute intervals at 350 or 400 m.p.m. A horse capable of recovering from this level of work in 10 minutes should be fit enough for novice and intermediate one-day events, remembering that the latter are longer and therefore need greater fitness. Recovery within the time after five repetitions would indicate advanced one-day event fitness.

It cannot be overstressed that all horses are individuals and must be treated as such. The tables in following chapters are merely a guideline and do not allow for lost shoes, heavy going or lazy horses. The real skill in training lies in the ability to design programmes for each horse and to recognise the need to adapt the programme without hindering the horse's progress. The same programme may take up to two weeks longer with a different horse. Eventually, the horse is working three 5 minute trots at 220 m.p.m. (3.3 km or approximately 2 miles); and three 3 minute canters at 350 m.p.m. and two at 400 m.p.m. (5.5 km or approximately 3.5 miles in total).

This sort of training should not be carried out more than twice a week, and if a horse is experiencing difficulty there are several factors that can be changed to suit the individual animal:

- distance of each effort
- speed of each effort
- interval between efforts
- repetition of consecutive efforts with intervals between

It must also be remembered that the training area will vary as the weather varies, and the trainer must be prepared to alter the training programme to allow for this.

Interval Training to More Advanced Levels

The demands made on a horse competing in three-day events, National Hunt racing and point-to-pointing require that the horse is at peak physical fitness and condition. The horse trained to the level previously described should be capable of working aerobically during canters of up to 500 m.p.m. Cantering for longer times does not actually increase fitness and thus speed and/or hill work must be used to train horses at these advanced levels. This is particularly true for part-bred horses.

For the six weeks prior to a three-day event, the speed and length of the canter workouts would be gradually increased so that the last minute was performed at steeplechase speed (690 m.p.m.), preparing the horse to go the required distance at that speed. The length of each gallop is thus kept down to 690 metres (less than half a mile) but the repetitions can be gradually increased up to five.

The final levels that different riders work up to vary. For example:

(1) Three repetitions of 9 minute canters at 550 m.p.m., 2 minutes' walking between each with the last minute of the final two canters being done at 690 m.p.m.
(2) Three repetitions of 10 minute canters at 550 m.p.m., 3 minutes walking plus one full-speed 800 metre (half-mile) gallop at the end.
(3) Some riders believe that three repetitions of 8 minute canters are adequate and that increasing to 10 minutes does not significantly increase fitness, provided the canters are fast enough.

These short distances at high speed develop speed and strength; the longer, slower canters develop strength, rhythm and staying power. If the speed of each effort (workout) is gradually and progressively increased, the speed of the horse is developed. As the horse adapts to the speed required over a given distance, the number of workouts can be increased to develop stamina at that speed. During racing and three-day eventing maximum muscle contraction is required and high lactic acid levels build up in the muscles. In attempting to have a more efficient way of dissipating high lactate levels, it is important that these levels of exercise are experienced during training – hence the introduction of a short gallop at the end of each canter workout.

Points to Remember

(1) Interval training must be monitored by pulse and respiration rate, *not* by time alone. It is the pulse rate immediately after the workout that will show you how much stress the horse has been subjected to, and the recovery after 10 minutes that will show you how fit the horse is. Without these results, interval training is meaningless and can be positively harmful.

(2) Pulse rate should exceed 200 per minute during the workout if *anaerobic* work is to be done. In the first minute after stopping exercise the pulse rate should fall to about 120 per minute if beneficial stress has occurred. A pulse of 150 indicates over stress and one of 100 shows that the horse has not worked hard enough.

(3) If pulse and respiration have not returned to normal within 30 minutes of completing the workout, the horse has been over-stressed and the programme should be adapted accordingly.

(4) The total of pulse and respiration per minute should drop by 30% in the 10 minutes after a workout.

(5) The respiration rate should never exceed the pulse rate. If it does, stop work.

(6) Never finish the workout if the horse appears distressed in any way, e.g. gasping, excessive blowing, stumbling or reluctant to continue.

(7) On the other hand, the horse must be stressed enough to stimulate the body systems to become better adapted for exercise. The pulse rate must be raised to between 80 and 150 per minute after a workout.

(8) It is possible to overtrain horses. If the horse is getting fit more quickly than expected, increase the distance of the workout rather than the speed. Speed is very stressful and long, slow work for a few days may be more beneficial.

(9) The interval training programme must be planned backwards from the proposed event(s) so that workout days fall into appropriate places before the competition.

(10) The horse may become bored. It is important to try to vary training schedules and to change the area over which training is taking place. Swimming is useful in maintaining aerobic capacity, particularly after minor lower limb injuries or for providing a

variation in the training programme.

(11) Always warm-up and cool-down horses properly on workout days, particularly if the horse has to travel in a lorry or trailer to the work area.

Interval training can also be adapted for training horses performing in long distance rides, and this will be discussed in more detail later in the book.

16 The Speedtest and Fartlek System

The success of training depends on the ability of the trainer to assess the horse's potential and to proceed accordingly, improving the horse's strong points and reducing its weak points. Breeding, conformation and 'feel' of a horse must all be taken into consideration. Dr Donald McMiken, working in America, has adapted some training methods used by human athletes for conditioning racehorses.

The Speedtest System

This system involves a three-stage programme. The horse is given a test at the end of each stage, hence the name 'speed test system'. The idea is to think of conditioning, or getting a horse fit, as a pyramid, starting at the base with slow work and, as the muscles and skeleton adapt, gradually increasing to fast work and competition at the peak. The broader the base established, the higher the peak that can be reached.

The First Three Months

A mature horse would spend the first three months of the training programme doing a lot of slow work at walk, trot and canter. After two weeks walking only, slow trotting would be incorporated so that the horse would be doing 3 km (2 miles) of mostly trotting. This would be built up so that every third day the horse would be doing 10 km (6 miles) of mostly slow canter work. The work on the other two days would be easier and shorter. The horse should be exercised five or six times a week. This part of the programme would have to be stretched if the horse were immature or had never been got fit before. Once an older horse has been fit and then let down, it is much easier to get fit again.

The aim of this slow work is to strengthen the bones, muscles, tendons and ligaments. Different tissues grow stronger at different rates; bones and tendons adapt to work more slowly than muscle, hence the long period of slow work.

It is felt to be important not to relentlessly pile more and more work on the horse. To this end, the programme is worked in four-week cycles: a moderate week, a hard week, a moderate week and an easy week. This means that the horse's body has a chance to repair any damage that may be done during the hard week before it is asked to work again. In traditional training routines, the horse may appear to be getting stronger but the damage can be slowly building up.

Fartlek

Fartlek is a Swedish word that translates as 'speed play'. This is a term used by human athletes to describe a philosophy as a training method. Human runners run as the ground allows and as their bodies tell them, slowing down and speeding up as they feel appropriate. Thus the man is always running within his limits, alternating fast and slow work in an essentially unplanned manner.

Fartlek is used to describe gallops the horse is given towards the end of the three-month period. These are short gallops, 200 to 600 metres (200 to 600 yards).

Obviously, translating fartlek to horse work is difficult – the horse cannot tell you how it feels. This is where the use of a heart rate monitor comes in. Electrodes in the girth collect the heart rate infomation and transmit it to the wrist monitor, giving a continuous readout so that the rider knows how stressed the horse is all the time.

Heart rate monitors muscular effort and energy production. Race-horses are given gallops of less than 1 minute and their heart rate is pushed up to 200 to 210 beats per minute. An eventer would probably only need to go to 190 to 200 beats per minute. For gallops of longer than 1 minute (as occur later in the programme), the heart rate should not be allowed to go any higher than 180. If the heart rate goes too high, the horse is slowed slightly so that it is never being overstressed.

Horses are never galloped more than twice a week, and not at all during easy weeks.

Speedtests

Speedtests are used to determine whether or not the horse is ready to move on to the second phase of the programme. This is a moderate gallop of 4 km (2.5 miles), during which the heart rate is recorded. The heart rate and speed are compared with the horse's previous efforts and with those of other horses in training. There is no set point at which a horse is said to be ready to move on, but a progression in heart rate indicates increasing fitness.

The Next Six Weeks

The second phase lasts about six weeks and is designed to strengthen the respiratory system and the muscles. The only way to do this is by stressing these systems, so the amount of fast work increases.

Racehorses are asked to gallop a mile or so at about 2 minutes to the mile (1 to 2 km at 800 m.p.m.). Hill work once a week is a good idea, as it sends the heart rate up at slower speeds and there is thus less wear and tear on the front legs.

The eventer would be asked to do a 2 to 3 mile (3 to 5 km) gallop, the speed being adjusted so that the heart rate never goes above 180. The earlier fartlek sessions have acted as ground work for this stage. Fast work takes place twice a week, the other days being similar to the first stage but speed is increased and distance reduced.

At the end of about six weeks, the horse undergoes another speedtest to show whether or not it is fit to continue to the next stage.

The Third Phase

During the third stage the conditions of the competition are simulated, the duration of work is reduced and the speed is increased.

For the racehorse, this means work at speed. This fast work is performed in intervals; the horse gallops, say, just under a kilometre (about half a mile) at racing speed, trots until the heart and respiration rates have recovered, and is then galloped again. The aim is to push the horse to near maximum to get the greatest fittening effects and yet to avoid fatigue (see the chapter on interval training). Three intervals at fast speeds twice a week are adequate. The distance of the gallops may be increased by the number of workouts is not changed. A final speedtest is given before the horse is raced.

The eventer would use speeds, distance and activities appropriate to its competition. For example, an advanced event horse would do three gallops of 1 kilometre (0.75 miles) at steeplechase speed. A novice event horse need gallop no faster than the speed needed in competition.

It is claimed that this method of training produces a much higher level of fitness than conventional methods, and this appears to be borne out by scientific studies. In Sweden, adult horses trained for only five weeks in this way developed greater muscle endurance (measured biochemically) than horses actually racing.

Horses involved in dressage and showjumping will not need the third stage of the programme, and some will not need the second as their early work will have strengthened them enough. Schooling for dressage and jumping can be fitted into the work of the first phase. Light workdays and easy weeks are good times to practice early dressage movements because these exercises are not too demanding physically. Jumping and advanced dressage are more stressful and fit into the moderate week's work. Thus not only are conditioning and training combined but the work pattern is varied and keeps the horse from being bored.

17   Monitoring Fitness

One of the biggest problems the trainer or rider faces during a horse's training programme is judging how much the horse can take. Every time the workload is increased, the trainer has to decide whether the horse can stand the work or is on the verge of injury or overstress. Although the horse's legs may be carefully examined for heat and swelling, by the time these external signs become apparent the damage may already be serious enough to interrupt the training programme.

In an effort to combat the 'trial and error' of getting horse's fit, there are more and more trainers (particularly in the U.S.A.) using technical equipment to assess their horses' fitness.

Thermography

This technique uses an infra-red camera to record changes in temperature. It can be used to spot wear and tear in the legs and feet where the inflammatory reaction has begun and the temperature of the area has increased but has not yet led to swelling and lameness.

This equipment is very expensive and is only used in a few specialised equine centres. It is possible to buy a hand-held sophisticated thermometer that, when skimmed along the leg about a centimetre above the surface, gives a digital readout accurate to 0.1 of a degree. The best results appear to be obtained using 'provocative cooling'. The legs are splashed with alcohol to cool them, and the thermometer measures which spot heats up the fastest.

Ultrasound

Ultrasound imaging uses ultrasound waves to produce detailed views of body tissue and is successfully used to detect pregnancy. Ultrasound bone analysis records the speed at which ultrasound waves pass through bone. There is a correlation between ultrasound readings and density of bone. As has been previously described, bone becomes more dense and strong with work, and ultrasound readings will reflect these changes in bone strength and also indicate the presence of damage. Bone under stress may suffer tiny microfractures that begin to build up and may lead to a serious fracture.

The equipment is expensive and the results are difficult to interpret. However, as expertise grows, ultrasound bone analysis may help to spot bone damage early and monitor the horse's progress in one aspect of training. It may prove particularly useful for two-year-old racehorses.

Gait Analysis

This technique involves filming horses at different paces and analysing their action frame by frame with the aid of a computer. The horse's gait is rated for efficiency and studied for oddities that might predispose to future unsoundness. The racehorse's gallop might also be compared step by step with the gallop of past champions.

Force Plates

The Swiss-designed Kägi gait analysis system is designed to provide the vet and the trainer with another pair of eyes. The horse is walked or trotted across a 40×1.20 metre track made of hard-wearing rubber, with a central plastic section which carries 160 sensors and can be used in any weather. The sensors can record 1790 details in 0.5 seconds, and the computer printouts are available in 10 minutes. Through the computer, the printout will record all the forces acting on each leg, both independently and jointly with the other legs. Professor Mueller of Zurich University has developed printout patterns of sound horses and horses with various forms of lameness, and comparison to these can help the vet diagnose lameness and the trainer assess soundness. However, interpretation of the results is not easy – and correct interpretation is, of course, the first essential requirement for using the system sensibly.

Lactic Acid Monitoring

A major goal of training is to increase the horse's tolerance to the lactic acid build-up incurred during fast anaerobic work. Lactic acid is now thought to have a role to play in predicting a horse's performance ability. In a group of untrained horses, individuals of greater natural ability will travel faster before they first produce lactic acid. As a horse gets fitter, it is able to go faster before lactic acid production starts. This suggests that (a) it may be possible to screen youngsters for lactic acid tolerance early in their training programme (b) lactic acid monitoring may be able to follow the effects of training and indicate a fall in exercise capacity due to injury or overtraining.

Lactic acid production is dependent on fast twitch low-oxidative muscle fibres. Lactic acid production increases as the horse's system reaches its oxygen-delivery capacity, and the level at which this build-up begins depends on the horse's fitness. This threshold of lactic acid build-up (acidosis) is called the 'anaerobic threshold' and indicates the horse's level of fitness.

As the horse gets fitter, the anaerobic threshold is delayed. If the threshold stops rising, the horse has reached its limits in terms of lactic acid tolerance and fast twitch muscle fibre use – in other words, it has reached the limit of its ability to gallop fast. If the threshold starts to fall, this indicates overtraining, illness or injury.

A step test can be conducted consisting of a series of workouts at slow, medium and medium-fast speeds. The horse is given jogging relief between tests to get the heart rate down to about 100 before it is asked to work again. This step test can be given under fairly consistent conditions and the horse's performance, in terms of lactic acid build-up, monitored throughout training. Blood samples are taken before training starts and after each of the three speeds in the workout. The blood is then analysed for the lactic acid content, using a sophisticated machine. Taking the blood sample at the same time after each of the three steps is vital, as the lactic acid level can drop quickly after exercise stops.

A graph of the lactic acid values and speed of exercise can be plotted which will demonstrate where the anaerobic threshold lies. Training a horse just above the lactic acid threshold helps achieve rapid progress in the horse's ability to perform fast work over a distance, and also appears to reduce the incidence of training and racing injuries.

Muscle Biopsy

This technique has been mentioned earlier. A small piece of muscle tissue is removed from the horse's rump (gluteus medius muscle) with a biospy needle, and is immediately frozen. The tissue is still 'alive', and is taken to the laboratory where it is thawed, stained and evaluated under the microscope. All horses have a mixture of various fibre types, and the biopsy is designed to show what proportion the fibres are in, how big the fibres of each type are, what their aerobic potential is and how much fuel is stored in them.

A horse which has a high percentage of slow twitch fibres (type I) can excel in endurance work but will always be at a disadvantage when bursts of speed are needed. A horse with a high proportion of fast twitch low-oxidative (type IIB) is a natural sprinter. Fast twitch high-oxidative (type IIA) are the most versatile, able to increase both power and endurance. Thus a knowledge of a horse's muscle fibre type can help assess its potential for various sports.

Heart Rate Monitors

These have also been discussed previously. Monitoring a horse's rate is becoming increasingly popular as a way of assessing fitness and monitoring the success of a training programme.

Videos

Getting a horse's performance on videotape can be very useful in monitoring its progress. The horse's movement and action can be watched in slow motion; the tape can be stopped so that every stride and movement can be analysed. If a horse is not jumping properly or a dressage horse is not performing movements correctly, the training programme can be adjusted to include the necessary specific exercises.

Computers

These machines can help keep track of all the variables that may affect a horse's training programme – for example, the horse's health, the condition of the going, the weather, etc. A computer programme can be used which reminds the trainer of factors that will determine how much work the horse should receive: soundness, history, most recent work and reaction to it, level of competition, and so on. With the appropriate information fed into it the computer can come up with a description of what work should be next in the training programme.

A computer may not be vital to success, but it can act as a valuable tool, preventing the trainer from overlooking factors that may affect soundness and performance.

18   Training Methods

Training the Event Horse

There has probably been more written about getting event horses fit than those for any other sport. As with any performance horse, the ideal degree of fitness at any time depends on the stage in the total training plan, the ability of the rider and the realistic competitive goals. The novice horse would be got to a similar standard of fitness as the hunter, while the advanced horse would require a fitness standard similar to that of the point-to-point horse. Indeed, traditional methods of training eventers follow very similar lines to those given for hunters and pointers. Obviously, schooling in the three disciplines of dressage, showjumping and cross-country would have to be incorporated in the training programme.

Getting the three-day event horse ready is another matter. This horse has to be supremely fit and yet disciplined enough to perform a demanding dressage test. The fitness programme has to ensure that the horse achieves a maximum level of fitness while staying sound and keeping a reasonably calm frame of mind. Interval training, which has been described in chapter 15, is used by some riders, while others adhere to more traditional methods.

In terms of body condition eventers lie midway between racehorses and showjumpers or dressage horses. The racehorse needs a powerful heart, big lungs and plenty of muscle, all operating at maximum efficiency, and not an ounce of spare weight. Most trainers consider that each racehorse has an ideal racing weight. The dressage horse, if wound up tight as the 'lean, mean machine', would never stand the mental stress of training; the aim is closer to that of the ball-room dancer – light on the feet, beautifully controlled, calm and confident, but often not a sylph-like figure.

Like the show horse, although not to the same degree, the dressage horse has to have a good top line to be pleasing to the eye; also, it needs extra muscle to carry the rider's weight further back and to produce collection. The showjumper, like the racehorse, probably has an ideal weight; it has to be powerful – which needs plenty of muscle – but never ponderous.

The event horse, being the master of many trades – distance, speed, power and control – has to compromise. It must be fast enough to do the cross-country in the required time without distress, and yet combine the other talents and physical attributes. Thus it will not be as lean and wiry as the racehorse, though less well-covered than the dressage horse. It must avoid the strain of extra weight on the limbs and the wind. However, there is muscle development of the neck and back, achieved by schooling. The major changes in the eventer's physique are due to fitness.

The event horse has to be taught to gallop in a balanced and calm manner; it must be able to lengthen and shorten its stride and cope with gradients when carrying the rider's weight. Some types of horses may find it difficult to gallop – the stuffy, short-striding horse is a prime example, and this type may need to be worked more strongly than a big, long-striding horse. The latter may need working more slowly to keep it calm and balanced. A reluctant or lazy horse may benefit from being galloped alongside another horse which will teach it to enjoy galloping and to try harder.

The Novice Eventer

The one principle that all authorities agree on is that the horse must be brought into work slowly and the first month of work must consist of walk, trot and slow canter on good going. Ideally, the programme outlined for the hunter should be used: 2 weeks of walking work followed by 2 weeks of walking and trotting work. After trot-work has been introduced, i.e. week 3, flat work in large circles and on a good surface can be introduced. With the novice horse, flat work should be part of the daily routine, consisting of no more than 30 minutes before or after about 1 hours' road work. As the horse gets fitter, lunge work can be incorporated two or three times a week – 20 minutes would probably be sufficient. Work on the lunge in side-reins can be very strenuous for the young horse and should not be overused until the horse is fitter.

Canter-work can be introduced in week 4 or 5. If possible, it is a good idea to enter one or two dressage competitions at this stage. This gives the novice horse a few outings to accustom it to travel, crowds, dressage boards and other monsters!

During week 5 or 6, jumping can be introduced into the programme, starting with pole work, grid work and small jumps in the school. This can be incorporated into the schooling programme so that the horse is jumping two or three times a week. By the end of the second month, the novice should have done a small local showjumping class or two.

Cross-country jumping must also be practised, and by the end of week 8 of the training programme the novice should either complete a couple of hunter trials or be schooled over cross-country jumps.

Midway through the third month of training (ten weeks), the novice should compete in a one-day event. The aim should be to ride a calm, accurate dressage test and to go clear, cross-country and in the showjumping. The intention is for the horse to clearly view and understand the cross-country fences and to be happy in its work.

The rider of the young horse who blazes round with the horse's head up may have gone fast enough to win but he may have spoilt a good young horse in the process. The object of this event and indeed the first few events, is to teach the horse about its job. These early events should be considered part of the training programme.

Once the novice horse is fit, it can compete in a one-day event twice every three weeks or once a fortnight. It is important that there is enough time between competitions to iron out any problems that may have arisen during competition. After about three months of competition, the young horse should be given a short break and let-down a little; this will ensure that the horse does not become stale or jaded. The length of this 'holiday' will vary between horses: two weeks out at grass by day and in at night with perhaps 20 minutes lunging a day may be enough.

After this break, the horse will need two weeks' training and schooling before competing again. By now, the rider should be aiming to get some points in the competitions – depending, of course, on how the horse is progressing. After another three to four months of competition, the horse should be roughed off completely for two months. It is important to corn feed the horse throughout this period so that it does not lose condition.

This means that the year is split up as follows:

January to March:	getting fit
April, May, June:	competing
Early July:	rest
Late July:	regaining fitness
August, September, October:	competing
November, December:	roughed off

The months given are obviously flexible, depending on the competitions that the horse is entered for.

The schedules in this chapter are only intended as guidelines and must be adapted for the individual horse.

Table 18.1 Interval training to one-day event fitness: a 10 week schedule in detail

---

Week 4

Day 1: 60 minute hack including three 5 minute trots; check recovery

Day 2: Dressage plus 60 minute walk

Day 3: Dressage plus cavaletti schooling, 30 minutes plus a walk to cool down or turn out to relax

Day 4: Dressage plus 60 minute walk

Day 5: 60 minute hack including three 5 minute trots; check recovery

Day 6: Dressage and cavaletti schooling, 30 minutes plus a walk to cool down or turn out to relax

Day 7: Rest day – turn out for exercise

Week 5

Day 1: Dressage plus 90 minute hack including 15 minutes' hill-trotting

Day 2: Dressage plus 60 minute hack including three 5 minute trots and one 4 minute canter (400 m.p.m.); check recovery – if the pulse has not dropped by at least half in 10 minutes, stay at this level and do not increase work

Day 3: Dressage plus 60 minute hack

Day 4: Dressage plus cavaletti schooling, total 45 minutes; then walk cool or turn out to relax

Day 5: Dressage plus 90 minute hack

Day 6: 30 minute hack plus three 5 minute trots and one 6 minute canter (400 m.p.m.); check recovery

Day 7: Rest day – turn out for exercise

Week 6

Day 1: Dressage plus 60 minute hack

Day 2: Dressage and jumping schooling; walk cool and turn out to relax

Day 3: 30 minute hack plus three 5 minute trots and two 4 minute canters at 400 m.p.m.; check recovery

Day 4: Dressage plus 60-minute walk

Day 5: Dressage plus jumping schooling; walk cool and turn out to relax

Day 6: 30 minute hack plus three 5 minute trots, two 4 minute canters (400 m.p.m.) and one 6 minute canter (400 m.p.m.)

Day 7: Rest day – turn out for exercise

Table 18.1 continued

Week 7

Day 1: 90 minute hack

Day 2: Dressage schooling

Day 3: 30 minute hack plus three 5 minute trots and three 4 minute canters (400 m.p.m.)

Day 4: Jumping schooling; walk cool and turn out to relax

Day 5: 90 minute hack

Day 6: 30 minute hack plus three 5 minute trots and three 4 minute canters (400 m.p.m.)

Day 7: Rest day – turn out for exercise

Week 8

Day 1: 2 hour walk

Day 2: Dressage schooling

Day 3: 30 minute walk plus three 5 minute trots, two 4 minute canters (400 m.p.m.) and one 6 minute canter working up to 500 m.p.m. for last minute

Day 4: Dressage plus 90 minute hack

Day 5: Dressage and jumping schooling; walk cool and turn out to relax

Day 6: As day 3

Day 7: Rest day – turn out for exercise

Week 9

Day 1: 2 hour hack

Day 2: Dressage and jumping schooling

Day 3: 30 minute hack plus three 5 minute trots, three 4 minute canters and one 6 minute canter (400 m.p.m.)

Day 4: Dressage plus 60 minute hack

Day 5: Dressage and jumping schooling; walk cool and turn out to relax

Day 6: 30 minute walk plus three 5 minute trots, three 4 minute canters and two 6 minute canters (400 m.p.m.)

Day 7: Rest day

Week 10

Day 1: Dressage plus 60 minute hack

Day 2: Dressage and jumping schooling

Day 3: 30 minute hack plus three 5 minute trots, three 4 minute canters (400 m.p.m.) and two 6 minute canters (450 m.p.m.)

Day 4: 2 hour walk

Day 5: Dressage schooling; turnout just to relax (not too much grazing)

Table 18.1 continued

Day 6: First horse trials
Day 7: Rest day – turn out to unwind

All rest intervals are 3 minutes walking.

Table 18.2 Interval training to one-day event fitness: a 10 week schedule in outline

The rest interval between work periods is 3 minutes in all cases. Only workout days are shown; 60 to 90 minutes hacking and schooling would be given on other days, with the horse given one rest day a week and turned out in the field to relax whenever possible.

Week	
4	1.49, 1.49, 1.49, 1.49
	3, 3, *1.4*, *1.4*
5	3, 3, *1.4*, *1.4*
	3, 3, 3, *3*
6	3, 3, 3, *3*, *3*
	4, 4, 4, *3*, *3*
7	4, 4, 4, *3*, *3*, *3*
	5, 5, 5, *3*, *3*, *3*
8	5, 5, 5, *3*, **3**, **3**
	5, 5, 5, *3*, **3**, **3**
9	5, 5, 5, *4*, **3**, **3**
	5, 5, 5, *3*, *3*, **3**, **3**
10	5, 5, 5, *4*, *4*, **3**, **3**
	5, 5, 5, *3*, *3*, *3*, **3**, **3**

Three canter intervals of *4*, *5* and **4** minutes, the first two at 350 m.p.m. and the last at 400 m.p.m., would build the horse up to the same degree of fitness.

Figures in plain type indicate trotting at 220 m.p.m., those in italic type cantering at 350 m.p.m. and those in bold type cantering at 400 m.p.m. Rest intervals are indicated by commas.

Table 18.3 Interval training to one-day event fitness: 11 week schedule

---

Week 1
  Day 1: 30 minute walk
  Day 2: 45 minute walk
  Day 3: 50 minute walk
  Day 3: As day 3
  Day 5: 60 minute walk
  Day 6: As day 5
  Day 7: Rest day

Week 2
  Day 1: 60 minute walk; one 3 minute trot
  Day 2: 60 minute walk; two 3 minute trots, 3 minutes' walk
         between trots
  Day 3: As day 2
  Day 4: As day 2
  Day 5: As day 2
  Day 6: As day 2
  Day 7: Rest day

Week 3
  Day 1: 60 minute walk; two 3 minute trots
  Day 1: As day 1
  Day 3: 60 minute walk; two 4 minute trots
  Day 4: 60 minute walk; three 3 minute trots
  Day 5: 60 minute walk; three 4 minute trots
  Day 6: 60 minute walk; three 4 minute trots, short canter
  Day 7: Rest

Week 4
  Day 1: 60 minute hack; three 4 minute trots, short canter
  Day 2: 60 minutes including four 3 minute trots; one 4 minute
         canter (350 m.p.m.)
  Day 3: 60 minutes including four 3 minute trots; two 3 minute
         canters (350 m.p.m.)
  Day 4: As day 2
  Day 5: As day 3
  Day 6: As day 3 including cavaletti and pole-work
  Day 7: Rest day

Week 5
  Day 1: 60 minutes including three 5 minute trots
  Day 2: Gymnastic jumping schooling

Table 18.3 continued

   Day 3: Dressage schooling
   Day 4: Gymnastic jumping
   Day 5: Dressage schooling
   Day 6: 60 minutes including three 5 minute trots and one 4 minute
           canter (350 m.p.m.)
   Day 7: Rest day

Week 6
   Day 1: 60 minutes hack
   Day 2: Jumping schooling
   Day 3: Dressage schooling
   Day 4: Jumping schooling
   Day 5: Dressage schooling
   Day 6: 60 minutes including three 5 minute trots and two 3 minute
           canters (400 m.p.m.)
   Day 7: Rest day

Week 7
   Day 1: Hack including four 5 minute trots
   Day 2: Dressage schooling
   Day 3: Hack
   Day 4: Jumping schooling
   Day 5: Dressage and hack
   Day 6: 60 minutes including three 5 minute trots and three 3 minute
           canters (400 m.p.m.)
   Day 7: Rest day

Week 8
   Day 1: Hack
   Day 2: Work out as day 6 of week 7
   Day 3: Hack and schooling
   Day 4: Jumping
   Day 5: Hack and schooling
   Day 6: Cross-country school including two 4 minute canters
   Day 7: Rest day

Week 9
   Day 1: Hack
   Day 2: 60 minutes including three 5 minute trots, one 4 minute and
           one 5 minute canter (400 m.p.m.)
   Day 3: Hack plus schooling
   Day 4: Showjumping schooling
   Day 5: Dressage
   Day 6: Cross-country school including three 3 minute canters

Table 18.3 continued

Day 7: Rest day

Week 10
    Day 1: Hack
    Day 2: Showjumping schooling
    Day 3: Hack
    Day 4: Cross-country school including 6 minute canter (400 m.p.m.)
    Day 5: Hack
    Day 6: Dressage
    Day 7: Rest day

Week 11
    Day 1: Hack
    Day 2: Practice dressage test
    Day 3: Three 5 minute trots, one 4 minute and one 6 minute canter
           (400 m.p.m.)
    Day 4: Hack
    Day 5: Schooling
    Day 6: ONE-DAY EVENT
    Day 7: Rest day

The principles shown in 18.1 of turning out to relax and unwind apply equally to all training plans.

The training programmes shown in tables 18.1, 18.2 and 18.3 all involve interval training, each building up to different levels of workout:

(1) Three 5 minute trots, three 4 minute canters (400 m.p.m.) and two 6 minute canters (450 m.p.m.)
(2) Three 5 minute trots, three 3 minute canters (350 m.p.m.) and two 3 minute canters (400 m.p.m.)
(3) Three 5 minute trots, one 4 minute and one 6 minute canter (400 m.p.m.)

These different levels suit different horses, depending on their age, temperament and type; but each training programme is considered suitable for getting a horse that is initially totally unfit to one-day event fitness. A young horse being got fit for the first time needs longer to obtain both the education and fitness required for competition. The following year, the horse could be brought to novice one-day event fitness in 8 weeks.

The Three-Day Event Horse

Before the rider attempts to get a horse to this level of fitness, he needs to know the horse intimately and have a thorough knowledge of its requirements, capabilities and limitations. The breeding or type of horse, its present condition and its previous fitness must all be considered when devising a horse's training programme.

It is generally accepted that four months is required to get a horse from soft condition to three-day event fitness. Assuming that Badminton is the spring competitive goal, this would mean getting your horse up around Christmas time. This may not be either desirable or even possible in terms of yard organisation around the Christmas holiday, and often horses are not brought into work until the beginning of January. If you have only 12 to 14 weeks to get a horse to this standard of fitness, several points must be remembered:

(1) The horse must be corn-fed at grass.
(2) If the weather is unpleasant or the ground wet, the horse should be stabled at night.
(3) The break must not be long; two months is enough to rest the horse while not losing too much fitness.
(4) If the horse has not been brought to three-day event fitness before, as much as 4 weeks may have to be added to the training programme.

There are other problems associated with bringing up a horse in January. Frequently the roads are icy, slippery and not suitable for road exercise on an excitable Thoroughbred-type horse. However, where possible, the first 2 weeks should be spent walking, initially for 30 minutes, building up to 90 minutes. Trot-work and schooling can be introduced and increased during weeks 3 and 4. Canter-work and small jumps can be introduced during weeks 5 and 6.

The first 6 weeks of gradually increasing levels of strengthening work will be basically the same in all yards. After this, the trainer will decide on whether to follow traditional training methods or interval training. Whichever routine is followed, it is very important to plan the programme carefully and monitor the horse's progress. Competitions, clinics and lessons must be incorporated, bearing in mind the major competitive goal.

The rider and trainer must be aware of the rigorous task that they are training the horse for. On cross-country day it will be required to complete four phases, consisting of:

Phase A: Roads and Tracks. Approximately 3850 metres (2.4 miles) covered at an optimum speed of 220 m.p.m. This is a good working trot or a mixture of walk and canter, aiming to cover 1 kilometre in 4 minutes.

Phase B: Steeplechase. Approximately 2760 metres (1.6 miles) covered at an optimum speed of 690 m.p.m. This is a fast gallop and the horse must be taught to jump safely at speed. If it is pushed out of its natural galloping stride it may fall, break down or become tired during phase D, so if the horse is not naturally fast it may be better to accept a few time faults.

Phase C: Roads and Tracks. Approximately 5940 metres (5.6 miles) to be ridden as Phase A.

Phase D: Cross-country. Approximately 5200 to 5700 metres (3.2 to 3.5 miles) to be ridden at 520 m.p.m. This is a fairly fast gallop with obstacles which must be safely negotiated.

This format gives the rider and trainer an idea of the distance and speed that the horse must be fit enough to cope with. The aim of any training programme, be it traditional or interval training, must be for a horse to be able to undergo this severe test and complete it at the required speed without fatigue in the latter stages – which could lead to a silly mistake and a fall. Also, the horse must still be fit enough to come out the next day not unduly stiff, and to show-jump accurately.

Traditional Methods

These methods involve long distances of relatively slow galloping to develop the horse's muscles, heart and lungs. A horse at one-day event fitness would be completing an 800 to 1200 metre (0.5 to 0.75 mile) gallop at half speed once a week. This would be gradually increased to 1200 metres (0.75 miles) at half-speed plus 800 metres (0.5 miles) at three-quarter speed, and then to 1600 metres (1 mile) at half speed and to 1600 metres (1 mile) at three-quarter speed by the end of the week 13. Prior to a three-day event, the horse should be able to gallop for 3000 metres (2 miles), starting at about 550 m.p.m. and finishing fast and on the bit.

The second piece of fast work to be completed during a week's programme would be longer, slower canter-work. Canter-work at about 400 m.p.m. would be built up over the 4 weeks prior to the competition to 20 minutes' sustained work. This method is popular for getting point-to-point horses fit but can put strain on the horse's legs. Care must be taken to observe any signs of heat and swelling, and fast work must not be repeated until the horse's legs return to normal.

Interval Training

Tables 18.4 and 18.5 illustrate two variations on the theme of interval training.

Table 18.4 Interval training up to three-day event fitness: 16 week schedule

---

Week 6
    Day 2: Three 3 minute canters (400 m.p.m.)
    Day 6: Three 3 minute canters (400 m.p.m.)

Week 7
    Day 2: Three 4 minute canters (400 m.p.m.)
    Day 6: Three 4 minute canters (400 m.p.m.)

Week 8
    Day 2: Two 4 minute and one 5 minute canters (400 m.p.m.)
    Day 5: Cross-country school
    Day 6: Walking exercise

Week 9
    Day 2: Cross-country school
    Day 6: Three 5 minute canters (400 m.p.m.)

Week 10
    Day 1: One 5 minute and two 6 minute canters (400 m.p.m.)
    Day 3: Three 5 minute canters, last minute at 600 m.p.m.
    Day 5: Showjumping
    Day 6: ONE-DAY-EVENT

Week 11
    Day 6: Three 5 minute canters (400 m.p.m.)

Week 12
    Day 2: Three 6 minute canters (400 m.p.m.)
    Day 6: Three 6 minute canters, last minute at 640 m.p.m.

Week 13
    Day 3: Three 6 minute canters, last minute at 640 m.p.m.
    Day 4: Walk
    Day 5: Showjumping
    Day 6: TWO-DAY EVENT (including steeplechase)

Week 14
    Day 6: Three 6 minute canters at 400 m.p.m.

Week 15
    Day 2: Three 6 minute canters, last minute at 640 m.p.m.
    Day 6: Three 6 minute canters, last minute at 640 m.p.m.

Table 18.4 continued

Week 16

   Day 1: Three 6 minute canters, last minute at 640 m.p.m.

   Day 3: Walk

   Day 4: Veterinary check for three-day event

   Day 5: Dressage

   Day 6: Cross-country } THREE-DAY EVENT

   Day 7: Showjumping

Table 18.5 Interval training up to three-day event fitness: 15 week schedule

---

Week 7

Day 2:	30 minute hack plus three 5 minute trots, two 4 minute canters (400 m.p.m.), one 8 minute canter, one 6 minute canter
Day 6:	30 minute hack, three 5 minute trots, two 4 minute canters (400 m.p.m.); steeplechase practice – 400 metres at 400 m.p.m. 800 metres at 640 m.p.m. (1 minute 15 seconds); slow down over 400 metres

Week 8

Day 2:	30 minute hack plus three 5 minute trots, three 4 minute canters, one 6 minute canter (400 m.p.m.) and one 4 minute canter at 520 m.p.m.
Day 4:	Cross-country school
Day 6:	Dressage or showjumping show or workout as day 2

Week 9

Day 2:	30 minute hack plus three 5 minute trots, two 4 minute canters (400 m.p.m.), one 6 minute canter (520 m.p.m.)
Day 6:	ONE-DAY EVENT

Week 10

Day 6:	Three 5 minute trots, two 4 minute canters (400 m.p.m.), two 6 minute canters (520 m.p.m.)

Week 11

Day 2:	Three 5 minute trots, two 4 minute canters (400 m.p.m.), two 6 minute canters (520 m.p.m.)
Day 6:	Show or workout as above

Table 18.5 continued

Week 12

Day 2:   Three 5 minute trots, two 4 minute canters (400 m.p.m.), school for 1600 metres at steeplechase speed (640 m.p.m.); pull up over 800 metres

Week 13

Day 5 or 6: Three 5 minute trots, two 4 minute canters (400 m.p.m.), two 6 minute canters (520 m.p.m.)

Week 14

Day 3:   Three 5 minute trots, two 4 minute canters (400 m.p.m.), one 6 minute canter (520 m.p.m.), one 1.5 minute gallop (640 m.p.m.), one 4 minute canter (400 m.p.m.)

Day 6:   Showjumping schooling

Week 15

Day 4 ⎫
Day 5 ⎬ THREE-DAY EVENT
Day 6 ⎭

First 6 weeks as for one-day event training; unless stated otherwise, exercise is 60 to 90 minutes' walking exercise plus dressage schooling, with the horse being turned out in the field to relax whenever possible.

Interval training is very flexible, and canter speeds, duration and repetition can be altered to suit each individual horse. These tables are merely guidelines, and each trainer/rider should plan his own training programme. The fast work days should be put in first, including actual competitions and cross-country schools, and then the slow work days fitted in.

If the horse has a setback, the programme must be replanned. If the horse does not feel right, the programme may have to be altered. It is very important for the schedule to be flexible. Always remember not to overwork the horse, particularly in the last week before competition, so that the horse arrives at the event with a contented yet controllable outlook.

Training the Long-Distance Horse

Long-distance and endurance riding can cause severe distress in a horse that has not been properly prepared. Riders must be capable of training a horse to a very high level of fitness brought about by following a carefully prepared schedule of regular exercise. There are no short cuts when getting to this degree of fitness. The rider, too, must be fit, as a tired rider hinders the horse.

The rider must also be able to recognise subtle signs of stress in the horse and be able to take suitable action to prevent any further distress. The long distance horse can suffer from muscular, cardiovascular and metabolic stress.

Muscular stress is seen when a tired horse starts to flag and stumble and muscle tremors may be seen in flank and thigh. The pace should steadied or the horse will become exhausted.

Cardiovascular stress can be monitored by checking pulse and respiration rates. Indeed, long distance riders must be very aware of pulse and respiration rates as horses are not allowed to continue until these values are considered suitable by the vet in charge. The normal ratio of pulse to respiration is between 2:1 and 3:1. If the cardiovascular system is stressed the respiration rate may increase dramatically, equalling that of the heart. A heart to breathing ratio of less than 2:1 is serious and the horse must be rested immediately.

Metabolic stress may manifest itself as heat stroke, dehydration, azoturia or colic, which have been discussed previously.

The Training Programme

The routine will vary according to the type of event entered for. A hunting fit horse would be able to complete a short ride of 30 kilometres with no ill effects, but the hunter is not conditioned to maintain a steady pace for prolonged distances.

Assuming that the horse is sufficiently fit for light hacking, the following time should be allowed to get the horse fit for different events:

Plus 2 weeks	20 mile pleasure ride
Plus 4 weeks	30 miles at 6 m.p.h.
Plus 8 weeks	40 miles at 6.5 to 7 m.p.h.
Plus 3 months	50 mile endurance ride
Plus 4 months	Bronze or silver Golden Horseshoe ride
Plus 4½ months	75 mile ride/Gold in Golden Horseshoe ride
Plus 5 months	100 mile ride

The long-distance horse will benefit from a longer period of slow conditioning work than, for instance, the hunter. Some trainers advocate a two month period of walking and trotting work, beginning with 1 hour per day building up to 3 hours per day. Cantering work is introduced only when stamina and muscle strength are well developed. Many riders who participate in long-distance events do not have 3 hours a day to devote to exercising their horses. While it is preferable to exercise every day, it is perfectly feasible to get a horse fit riding it every other day, provided longer rides are given at weekends and the horse is got out for some exercise daily.

The horse should be ridden over varied terrain. This helps prevent the horse from becoming bored and also helps build up muscle. However, hills are not necessary for getting a horse fit provided adequate time and distance is put in.

The horse must be made to work during these rides – this will help to develop its paces. If the horse has a natural slow, short-striding walk, this should be improved until the horse is able to average 4 or 5 m.p.h. walk. During a ride the horse must be relaxed in walk because it is the pace used for recovering and 'resting'. The rider must aim for a long, free-striding walk on a light contact, with the horse calmly picking its way over the ground.

In actual competition, trot is mainly used. It is a pace that covers the ground, spreads the workload evenly, and is thus less tiring for the horse. The rider must remember to change the diagonal frequently. Again, the trot must be encouraged to be long-striding and easy; a good, strong trot will average 9 to 10 m.p.h., a slow trot about 6 to 7 m.p.h. Each horse has a speed within a pace at which it is happiest working, and forcing a horse to go faster is tiring for both horse and rider. If a horse has a naturally slow trot that has not been improved by training, it should not be hurried and instead time should be made up by cantering.

A rolling canter averages about 10 to 12 m.p.h. It is a useful pace in endurance rides where the horse needs to maintain canter for long distances, developing a machine-like rhythm.

Two different types of work can be used side by side in the training programme:

- shorter, sharper rides to develop the horse's wind
- longer, steadier rides to develop stamina and the steady, rhythmic paces necessary for successful competition

The lengths and speeds of rides during training will vary depending on the competition goal, and will be outlined in more detail later.

In long-distance work, almost more than any other equestrian sport, it is essential for the rider to know the horse. It is not a sport where one person can train a horse and another person successfully compete on it. Where horse and rider are a team, there is a greater likelihood of success.

Pulse and Respiration Rate

Long distance rides are subject to stringent veterinary checks, and pulse and respiration are among the checks made. This has resulted in the rider being very aware of his horse's pulse and respiration rates, using them as a guide of fitness during training and competition. The following points are useful to remember:

(1) Establish normal resting values.
(2) Establish unfit working values. Find a hill or convenient stopping point on one of the exercise routes, and record the rates after the horse has walked to the top. Walk on quietly for 10 minutes, and take them again. Take pulse and respiration rates at the same place each week, noting speed, climate and any other factors that might affect the horse's pulse and respiration.
(3) As the horse gets fitter, the initial readings will get lower and the horse will have recovered in the 10 minute interval. This indicates that the horse is ready for more strenuous work. As a rough guide, the horse can begin to work harder if the pulse is below 80 beats per minute after a work session. Allow the pulse to drop to about 60 before continuing work.
(4) The horse's pulse and respiration should have returned to normal within 30 minutes of finishing work. If, during a competition, the pulse has not fallen to below 70 after a compulsory rest period, the horse may not be allowed to continue.

Pleasure Rides

These are rides of 20 to 30 miles (30 to 50 km) length, performed at speeds of 5 to 6 m.p.h. (8 to 10 km/h). A horse that is hunting fit would be able to complete one of these rides with no ill effects. A horse that was hacking 6 to 10 miles (10 to 16 km), four times a week, would be adequately fit for a 20 mile (30 km) ride. A 30 mile (50 km) ride would require longer training rides including several miles of trotting at 8 m.p.h. (13 km/h) and canter-work. It would take about four weeks to get a horse from light hacking fitness to 30 miles at 6 m.p.h. (50 km at 10 km/h) fitness. The general routine would involve three days of ordinary exercise and two days with longer rides. The horse could be given sharp canters of about 1 mile (1.5 km) during the shorter rides, but the average speed of a ride should not be more than 10 to 12 m.p.h. (16 to 20 km/h) because this imposes too much stress on the horse.

A week's exercise may be along the following lines:

Monday: 9 km (6 miles) in one hour at a steady pace.
Tuesday: Semi-rest day; light exercise or turned out to grass.
Wednesday: 90 minutes at 10 km/h (6.5 m.p.h.).
Thursday: 2 hours including 3 km (2 miles) canter-work.
Friday: Semi-rest day.
Saturday: 3 hours at 8 km/h (5 m.p.h.).
Sunday: 30 to 35 km (20 to 25 miles) at a steady pace.

This routine makes best use of the weekends so that the rider who works full-time is still able to get his horse fit. On semi-rest days the horse is usually turned out in the field, depending on the time of year and the weather. Lunge work can be substituted for 20 to 30 minutes of the ridden work, provided the horse is lunged correctly.

Competitive Trail Rides (C.T.R.)

These rides vary in severity from 25 miles (about 40 km) in one day, through 60 miles (about 95 km) in two days, to 100 miles (about 160 km) in three days, and are judged on a 'time plus condition' basis. There is an optimum time, with penalties given for any loss in condition of the horse.

An extra four weeks should be allowed to get a horse from 30 miles (50 km) in one day to 40 miles (65 km) in one day fitness. The speed is usually about 6.5 to 7 m.p.h. (10 to 11 km/h) and only a fit horse can do this without stress. The horse must work six days a week over greater distance and at faster speeds. The programme for the last four weeks before a 40 mile (65 km) competitive trail ride at 7 m.p.h. (11 km/h) could be as follows:

Week 1
Monday: 1.5 hours at 6.5 m.p.h. (10 km/h).
Tuesday: 1.5 hours at 6.5 m.p.h. (10 km/h).
Wednesday: 1 hour at 10 m.p.h. (16 km/h).
Thursday: 2.5 hours at 6 m.p.h. (9 to 10 km/h).
Friday: 20 to 25 miles at 6 m.p.h. (30 to 40 km at 9 to 10 km/h).
Saturday: 1.5 hours easy ride.
Sunday: Rest day.

Week 2
All distances increased by two or three miles (3 to 5 km).
Friday: 25 miles at 7 m.p.h. (40 km at 11 km/h).

Week 3
Three consecutive workdays consisting of one ride of 25 miles at 7

m.p.h. (40 km at 11 km/h) and two rides of 20 miles at 7 m.p.h. (30 km at 11 km/h).

Two days of light hacking and two days of walking work.

Week 4

The horse is just kept ticking over before the ride with a 20 mile (30 km) ride in 2.5 hours in the middle of the week.

To compete at this level is very time consuming, as the rider must be prepared to have a horse capable of doing more than the stipulated minimum requirement. The ultimate aim is to compete in 100-mile competitions, such as the Golden Horseshoe Ride. The horse must have a thorough programme of slow work as a broad base on which to build this sort of fitness. The horse must also be used to carrying the rider's weight for a long time. Another four weeks should be allowed before a horse would be expected to complete a Golden Horseshoe Ride with a Bronze or Silver award. Yet another two weeks should be added to this before a horse could be expected to gain a Gold award.

Endurance Rides

These rides vary in length from 50 to 100 miles (about 80 to 160 km) and the fastest fit horse wins. This means that the emphasis is on accelerated performance for long periods, so the training programme has to be conducted at greater consistent speed. It cannot be overemphasised that to keep a horse sound under this degree of stress there must be *a firm basis of long slow work*.

In order to be up with the leaders in a 50 mile endurance ride, the horse must average 10 m.p.h. (16 km/h). The initial training is similar to that for a competitive trail ride but the last four weeks concentrate at higher speeds, e.g. 20 to 25 mile rides at 10 m.p.h. (30 to 40 km at 16 km/h).

Endurance rides of 75 or 100 miles (about 120 to 160 km) are very tough, and absolute fitness if vital. The horse will average 8 m.p.h. (13 km/h) and be under saddle for 12 to 20 hours. The lead-up to these rides should include at least two 50 to 60 mile (80 to 95 km) rides. This not only increases the horse's fitness but allows the rider to establish the horse's individual pattern for the ride.

Interval Training for Long-Distance Riding

It may appear to be a contradiction in terms to use Interval Training for long distance riding, but a certain amount of fast work does have a role to play. The long distance horse needs a high proportion of slow twitch muscle fibres; these have a high ability to use oxygen, and fatigue slowly. Arabs have an inherently high proportion of slow twitch fibres and excel at endurance work. However, the horse is born with a certain proportion of slow twitch to fast twitch fibre, and this is not changed with training.

What training *can* accomplish is to develop the aerobic capacity of the fast twitch fibres. Training increases the proportion of fast twitch high-oxidative fibres, thus the endurance ability of the horse is enhanced. This is done by increasing the anaerobic threshold (the point where lactic acid build-up begins) by including faster work in the training programme, Interval training for 30 to 45 minutes twice a week, getting the heart rate up to 160 beats per minute for a short time without overstressing the horse, will increase the aerobic capacity of the muscles. This fast work has the added advantage of stimulating the cardiovascular and respiratory systems (heart and wind). Long, slow work has little effect on these systems but does serve to build up muscle, increase resistance to fatigue and develop the glycogen sparing effect, and it also accustoms the horse to carrying the rider's weight.

Interval training has been used very successfully to train long-distance horses. Six to eight weeks of walk and trot are essential before canter-work begins. Intervals of slow canter can then be introduced, building up to three 12 minute canter intervals with 3 minutes walking in between, in the last month before, for example, the Golden Horseshoe Ride. The slow long-distance work is given on the other days of the week, as outlined previously.

Pleasure rides and qualifying rides act as a very important part of the fitness programme. They offer a guide to fitness, but they also ensure that the rider is achieving the required speeds – this is particularly useful for the rider training alone.

Long-distance riding has become more popular recently. Many of those participating originally bought a horse to hack, then decided to compete in distance rides and found the discipline and skill needed to get a horse fit, the degree of competitiveness and the companionship developed between horse and rider, very fulfilling. Long-distance riding does, however, demand fitness from horse and rider, and should only be undertaken if the horse is adequately prepared.

Training the Dressage Horse

The dressage horse is the ultimate specialist, and its training does not rely on fitness so much as the ability to perform certain movements with ease, elegance and accuracy. That is not to say that the dressage horse does not have to be fit; indeed, to compete throughout a long season of indoor and outdoor shows, the horse must be in top condition. This involves mental as well as physical condition, and the dressage horse must possess the correct attitude to its work; it must accept discipline and not 'boil' under pressure.

The dressage rider trains the horse to make it more gymnastic and athletic, and to help develop its natural ability so that it is able to work with increasing ease, balance and elegance. Key words for successful dressage are suppleness, power and control. During training, the horse becomes more collected; its centre of gravity moves backwards and the hind-legs become increasingly engaged. At the same time, the horse develops muscle tone and power; the top line becomes more exaggerated, with a crested neck and well developed hind-quarters. However, the heart and lungs are not extended to the same degree as in a horse that is required to gallop.

Muscle development takes, quite literally, years of work to acquire, but is easily lost if the horse is not in work. Several weeks not being ridden would require months of work to regain lost muscle tone. This means that the working year of the specialist dressage horse is unlike that of other competition horses in that it is not roughed-off and turned away for a long rest. Instead, it is given two or three short breaks, of no more than 2 to 3 weeks, during the year. During this break, the horse would probably be turned out by day and brought in at night and corn-fed. This freshens the horse up without it losing too much muscle.

If, for some reason, the horse has been out of work for more than 4 or 5 weeks, it should be walked and trotted on roads and tracks for 2 to 3 weeks before dressage training starts again. Theoretically, the type of fitness required for a dressage horse can be achieved entirely by working in the school. However, great care must be taken not to strain muscles through overworking the horse in a confined space before a correct degree of muscle tone has been established. Road work should nevertheless be active, with the horse working in the correct outline for short periods during the hack.

Once the horse is fit, it can be schooled or lunged for up to 40 minutes per day, but sessions of collected work must always be interspersed with work on a loose rein so that the horse is able to relax and rest muscles that may be tired from unaccustomed exertion.

The training programme must be varied to prevent the horse from becoming sour or tense. Quiet hacking on the road before, after or instead of schooling will help the horse to relax. Loose jumping and gymnastic jumping exercises will encourage the horse to use it himself and keep supple. A canter in a suitable field to improve the heart and wind will also give the horse a break from the demands of collected work.

The dressage horse is improved by systematic intensive work, and year by year the horse advances until it has reached the limits of its performance. During the first year of the horse's training, it is taught to move forwards freely on the lunge and under a rider. Lateral work is taught in the second year, and the horse progresses to flying changes, greater collection and better extension during the third year of training. Thus, provided the horse, rider and trainer have the ability, the horse could be working on piaffe and passage in its fourth year of training, i.e. as an eight-year-old.

Actually getting the dressage horse fit is one thing, but long-term success depends on correct, careful and skilled training.

Training the Showjumping Horse

Showjumping is a specialist sport, and much of the fitness programme consists of schooling the horse and improving its jumping technique. Training and fitness go together; thus the training programme will vary according to the age and ability of the horse, and how much of the year the horse is competing.

Very little change is made to the natural shape of the showjumping horse. There is muscle development, but this is not as pronounced as in the dressage horse. Fitness and stamina are improved, but not to the same extent as in the event horse. Possibly the greatest change made to the showjumper is in its movement, particularly the canter which becomes more powerful, balanced and responsive. The aim of training is to enable the horse to make the best use of its natural ability to jump high and clean. The trainer does this by using basic flat work to develop balance and collection, encouraging the horse to have a rhythmic pace and to accept the bit. Schooling over jumps does not change the horse's ability, but does build up its confidence in itself and in its rider.

The showjumping circuit has developed into a highly professional and commercial world. There are two seasons that occupy the whole year. The summer season lasts from May to October, culminating in the Horse of the Year Show, and consists mainly of outdoor shows. The winter season extends from October to April, and competitions are usually indoors. the temptation is for riders to overwork their horses, but showjumping horses need a spell off work as much as any other competing horse.

The performance of the horse and the number of seasons of tough competition that a horse will be able to withstand depends on a correct combination of schooling, fitness, work and rest. Properly treated, a showjumper can compete at the top level well into its late teens. A horse will also last longer if it is not competed too early, before it is ready both physically and mentally. Plenty of time must be allowed to get the horse fit. If a horse is fittened too quickly, it can be detrimental to the time that it can stay fit, and this can affect not just one season but all the horse's competitive life. As with all disciplines, true fitness is based on a firm foundation of long, slow work.

Different trainers have different ideas on how long a horse should stay in work:

(1) 6 weeks semi-rest, 6 weeks fittening work and 3 months competition twice a year.
(2) 3 months rest, 8 weeks fittening work and 6 months competition. A long season like this would be broken by two short semi-rest periods of up to two weeks. The number, timing and duration of these short breaks would depend on the individual horse.

The less rest the horse is given, the quicker it can come back to competition fitness. However, the first 2 weeks of any fittening programme should consist of walking work on the road. The next 2 weeks should involve walking, trotting and basic ground work. Canter-work, more intensive schooling and jumping exercises would be introduced no earlier than 4 weeks after the horse has been brought up from its rest. The horse must be walked for 15 to 20 minutes before a schooling session to give it adequate time to warm up. An older, experienced horse will need less schooling than a young horse, and the seasoned campaigner would probably do very little jumping at home. The young horse, however, would need more schooling so that it arrives at the competition supple and obedient, and this may involve jumping two or three times a week. Gymnastic jumping is useful for building both confidence and the correct muscles.

Like the dressage horse, it is possible to get the showjumper to a suitable degree of fitness by schooling in an arena, because the movements practised at home are the same as those performed in competition – unlike the racehorse or eventer, where a great deal more is asked in competition. This, however, would not be ideal; a horse can easily get bored and stale and start to develop evasions, so work must be varied including quiet hacks, canters on suitable ground and perhaps even a day's hunting.

From a scientific viewpoint, the type of training outlined here is ideal provided that the horse meets in training a threshold of stress similar to that which it is going to meet in competition. Canter-work must not be neglected, because stamina is important in an animal that may spend several long days each week away from home and may jump five or six rounds a day, including timed jump-offs which place great stress by requiring fast turns and changes of pace. The horse that is a speed specialist should be trained by giving faster work to develop its aerobic capacity, while the puissance horse should be trained for power in a similar way to the dressage horse.

Training the Racehorse

The Point-to-Pointer

The initial training of the point-to-pointer will be the same as for the hunter. The horse may be got fit for the opening meet of the season if being raced early in the point-to-point season, or may be got up slightly later if racing late in the season. The point-to-pointer has to qualify by hunting six to eight times during the hunting season with one recognised Hunt. Bearing in mind the treacherous conditions that can prevail after Christmas, it is wise to qualify the horse before the end of December. This way, the horse can do one day a week during the first two months of the season leaving a clear six weeks or more to prepare it for its first race.

Specific training for the forthcoming point-to-point season should begin about four weeks before the horse's first race, provided the horse is hunting fit, in good condition and well-muscled. Hunting fit is *not* racing fit – even a horse capable of hunting two days a week is not trained adequately for galloping for 3 miles (5 kilometres) over 4 ft 6 in (1.5 metre) fences. The hunter rarely gallops steadily for long distances; there are checks, changes of pace and a variety of obstacles. The hunter may have a great capacity for endurance, but its capacity for fast work needs to be developed.

In other words, the point-to-point horse has to be galloped during training, and a suitable place must be found. This is not to say that if the horse is racing over 3 miles (5 kilometres) it must be galloped for 3 miles during training – this will tire the horse rather than get it fit.

It is generally accepted that horses should do fast work no more than twice a week, but the duration and speed of the fast work varies according to different trainers. Two different training programmes are given here and show alternative ways of reaching the same goal.

The aim of the first programme is to keep the hard, muscled-up hunter fit by giving a long, slow workout four times and developing its gallop by fast work twice a week.

Monday: Walking, trotting and possibly a quiet canter (300–350 m.p.m.) for about 1.5 km (1 mile); 60 to 90 minutes total.

Tuesday: Walk, trotting and a steady canter (350 m.p.m.) for up to 3 km (2 miles), according to the type of horse.

Wednesday: Steady canter (350 m.p.m.) for 1.5 km (1 mile) 20 to 30 minutes walk and trot and then a 1 km (5 furlong) gallop (600 m.p.m.).

Thursday: An easy day; walk and trot for 60 to 90 minutes.

Friday: As Tuesday.

Saturday: As Wednesday.

Sunday: Rest day.

This programme is very traditional, using prolonged periods of slow canter work to build up the horse's capacity for endurance, and keeping fast work to the minimum to avoid wearing the horse out.

In the second programme, the horse is again galloped twice a week. Initially, the horse is asked to go at a fast, hunting canter (about 400 m.p.m.) for about 2.5 km (1.5 miles). This is gradually increased over the next month to 3 km (2 miles) at a medium gallop (about 500 m.p.m.). One day a week the horse is given a sprint up a hill. Moving fast uphill puts less strain on the horse's legs than galloping on the flat, but it does demand a lot of the heart and lungs. Horses should not be asked to gallop downhill during training, as this puts a lot of strain on the front legs and increases the risk of tendon breakdown.

As the horse gets fitter, a sprint is included in the gallop; the horse is asked to stretch out over the last kilometre of its 3 km gallop. The horse should not be pushed to its limits, and the rider should always feel that the horse has something in reserve.

The rest of the week, the horse's exercise would consist of 60 to 90 minutes hacking on roads and tracks, including as much uphill trotting as possible.

Thus the weekly programme would look something like this:

Monday: Walk and trot, 90 minutes.

Tuesday: Steady, progressive warm-up 3 km (2 miles) gallop at 500 m.p.m. followed by a steady cool off and wind-down.

Wednesday: As Monday.

Thursday: As Tuesday.

Friday: As Monday.

Saturday: Uphill sprint.

Sunday: Rest day.

The training programme of a point-to-point horse will also include some jump-schooling. If the horse has raced before, it will not be necessary to give him much jumping practice – a gallop over schooling fences a week or so before his first race should serve as an adequate reminder.

If the horse has not raced before, it should be introduced to point-to-point fences gradually. Asking a hunter to gallop over fences is asking it to do exactly the opposite of what it has been doing in the hunting field. The hunter is taught to take its fence at a steady pace and to jump big and clear. The point-to-pointer must throw caution to the winds, jumping out of a fast gallop, not 'dwelling' at the fences, and galloping on immediately upon landing.

In order to establish this confidence, it is best to start with small, solid hurdles, well sloped and no more than a metre high. These fences are easy, and very soon the horse will become familiar with them and will gallop on over them, judging the take-off distance with confidence. The horse can then progress to brush schooling fences. Again these should not be too big and forbidding – the horse must not be intimidated at this stage. Jumping with a more experienced horse can be helpful. The older horse can act as a lead initially and then jump alongside the novice. It is important that the young horse is not flustered by jumping in a crowd, although this may happen in a race.

Races are usually won by horses that jump smoothly and quickly. This can only be accomplished by a horse with confidence in its own jumping ability, brought about by sensible schooling.

The two programmes given above are typical of those used in practice. However, previous chapters may have suggested that traditional methods will soon have to incorporate modern ideas in order to stay successful.

National Hunt Racing

The hurdler and 'chaser may come up through the ranks of flat-racing horses or may be bred specifically for jumping. A horse that has raced on the flat will be easier to get fit than an untouched youngster. However, a great depth of fitness is needed for National Hunt races, particularly for the longer distance races. This means that some trainers put a greater emphasis on initial slow work, allowing up to 8 weeks walking and trotting before canter-work starts. The early routine is similar to that used for getting hunters fit and is the same in most yards, consisting of about 90 minutes of walk and trot-work, including hill work. There is great variation between yards in the cantering and galloping routine.

After the initial slow work, canter would be introduced on, perhaps, two days a week, each canter session being about 2.5 km (1.5 miles). This would be gradually increased to four work days a week, so that the weekly programme would be as follows:

Monday: 90 minutes of road work, walk and trot.

Tuesday: 2.5 km (1.5 miles) cantering (half speed).
Wednesday: Fast work.
Thursday: 30 minutes walking only.
Friday: Cantering (half speed).
Saturday: Fast work.
Sunday: Rest day.

As with any horses kept at peak fitness for prolonged periods, mental happiness is essential for success. To prevent the horse from getting bored or stale, exercising routes should be varied; the horse could be turned out for short periods in a paddock and other forms of work, such as swimming, could be used.

Facilities for galloping jumpers are important, but need not be as straight as for the relatively-immature flat-race horse. It is not necessary to gallop horses over the distance that they race over. If the horse is in good health and well prepared from its road work, it just needs to be topped up once a week by strong work and kept supple on two other days by half-speed canters over 2.5 kilometers. The skill of a good trainer lies not so much in getting the horse fit but in keeping it fit for a series of races. Once a horse is fit, it should be kept simmering and only brought to the boil on race days. This means that the jumping horse is rarely given a full-speed gallop during training; the horses are worked well up into the bridle (never on a loose rein) at a half-speed gallop, so that they are able to quicken readily if asked for the last 500 metres.

The horse's jumping ability has such a large influence on the outcome of a race that home trials of a horse's fitness are rare. Schooling over fences is obviously very important, and before being taught to jump the horse must be obedient to the aids and have done enough groundwork to be balanced and co-ordinated. The young horse should be taught to jump out of a trot or slow canter over small fences before being asked to jump schooling hurdles. Training the horse to jump loose in an indoor or outdoor arena is useful and gives the young horse confidence in its ability to jump, teaching it to make its own adjustments without the weight of the rider on its back.

Just as interval training and other modern methods can be used successfully to prepare event horses, so too can these methods be used to prepare 'chasers. In particular, blood profiles and optimum racing weights are commonly accepted. Heart-rate monitors are still regarded as being in the 'secret weapon' category, but they probably offer the greatest scope for improved performance, monitoring fitness, detecting subclinical disease and thus offering trainers improved results and improved economics.

Training the Flat-Race Horse

Flat racing is the most natural of all the equestrian sports. Horses are athletes designed to use their speed to escape predators. The racehorse has to be taught to accept the rider's weight and a certain amount of control, but the crucial factor determining success is the horse's fitness. The horse must be got correctly fit for certain races which are carefully selected. Races vary in distance from five to twenty furlongs (about 1000 to 4000 metres) and, like human runners, each horse has its own best distance. Horses tend to stay better as they get older; two-year-olds are not raced over distances greater than eight furlongs.

Training flat horses is complicated by the fact that as two, three and even four-year-olds these horses are still growing and maturing. The body systems of young horses are frequently unable to withstand the stress of training, leading to a high percentage of wastage in racing Thoroughbreds. The majority of horses bred to race on the flat are broken after the yearling sales in October, and so start work at eighteen months old.

The breaking programme will involve lunging and long-reining, perhaps even long-reining through starting stalls. This gets the young horse used to stalls early on, avoiding problems at a later date. After about 4 weeks of groundwork, the horse will be backed and ridden away at walk and trot for 3 to 4 weeks. The youngster will be introduced to canter work on the gallops from about New Year onwards, beginning over 2.5 furlongs (500 metres) and gradually working up to 4 furlongs (800 metres). As the horse learns to canter in a balanced way and becomes fitter, a second canter is introduced into the same training session.

The next step is for the two-year-olds to canter 'upsides', i.e. alongside each other. Canter-work should be preceded by 20 minutes' walking and some trotting, as the canter-work increases, the length of time spent trotting is reduced so that the horse is having about 75 minutes' exercise.

As the horse gets fitter, its 'swinging' canter can progress to a half-speed gallop and eventually a sharp gallop twice a week. Initially, the fast work should be over very short distances; these are gradually increased, depending on the race the horse is being prepared for. Just before the race, the fast work should be shortened and quickened to prepare the young horse for full racing speed.

Flat horses are galloping specialists, and their training regime is geared to ensure that muscle development is confined to those muscles used for galloping – any extras are considered unnecessary. This means that their work is mostly composed of galloping and cantering, interspersed with hill work, swimming and walking on a horse walker.

The longer a young horse is given to mature and muscle-up, the better. Many horses will not race as two-year-olds, and only make their track debut as three-year-olds. Others mature earlier and are capable of racing as two-year-olds. Developing muscle is accomplished by steady exercise gradually increasing in duration, walking, trotting and hill work. The youngster should not be cantered until it can trot steadily and in a straight line, using its hocks and shoulders. The jockey must be careful to ride on both diagonals at the trot so that muscle development over the back is equal and balanced. Initially, the young horse should be cantered slowly so that it can find its own balance before being asked to gallop.

Scientific knowledge suggests that racing involves maximum contraction of the muscles involved with galloping, thus a more strenuous component must be brought into the training programme. In order for the body to have a more efficient way of disposing of the high lactic acid concentrations built up during a race, it is important that these levels are experienced during training. A racehorse can be trained using a speed endurance schedule, which is similar to interval training but is based on repeating high intensity exercise over short distances. Exercise might consist of several maximum gallops over 400 metres, with intervals of trot between for the horse to recover partially. This type of training is rarely seen in racing yards and it is not known if a young horse could stand this sort of regime either physically or mentally; however, for the older flat-race horse it holds promise of improved performance.

It is interesting to note that, using such methods, human runners have improved their performance out of all recognition. Accepting that training the horse is a totally different science, it is still interesting to note that racing trainers tend to use the same methods as their grandfathers and to produce only the same speeds. It may be that the legs of the horse are the limiting factor, but it is possible that a trainer using new techniques and equipment will start to break equine speed records.

Training the Driving Horse

There are many variations in the sport of driving, from scurry to marathon, concours d'elegance to Hackneys. This means that the type of horse used and the degree of fitness required will vary enormously.

The basic training, after the horse has been brought up from a rest, does not vary greatly. There is an initial period of walking work, either under saddle or in long-reins, lasting about 35 to 45 minutes per day. The horse must be made to use itself in walk, alternately collecting and extending the walk. Lunging and work in harness can then be introduced, including some trotting work. Lunging can be gradually increased up to

45 minutes per day.

It is from here that the training programme will diverge, depending on the competitive goals and sub-goals. A 'single' horse being driven at an exhibition class could start the fittening programme in March, working up to 30 minutes on the lunge, and be ready for showing in May. The marathon team horses that are expected to pull a load of about a tonne for 13 to 20 km (8 to 12 miles) would need up to 45 minutes on the lunge to reach adequate fitness.

A horse would need three months' training to be fit to compete in a three-stage driving trial. A form of interval training can be used, as these horses need to be very fit; 3 minute trot intervals followed by 3 minute walking intervals can be used, the number depending on the horse's reaction to the work. Once the horses are trotting five 3 minute intervals with ease, the working time can be gradually increased. The rest interval should remain at 3 minutes, and exercise repeated until the horses are sufficiently stressed but not overstressed.

The next stage in the ridden horse would be to introduce canter-work, but fast work is obviously unsuitable for the driven horse and hill work should be included to increase fitness. It is useful to have a measured drive; the driver should know the ideal time for the drive at certain points along the route, and he should carry a watch. Gradually, as the horses get fitter, aim for this optimum time. This measured interval drive should be used on a four-day schedule. By the time the major competition goal arrives, the team should be able to go 1.5 times the distance of the cross-country phase without distress.

As the horse gets fitter, it may be necessary to adjust its tack to allow for muscle development.

The horses must work together as a team, and the driver must know his horses and recognise subtle signs of distress or abnormal behaviour. During the cross-country phase, it is possible for horses to become tired and even exhausted. They may still have the ability to go on but they lack the muscle control and mental ability to cope with a problem – for example, a hazard on the cross-country. Fatigue is probably the biggest obstacle of all. The horse's gait changes and the incidence of accidents and injuries increases. Fatigue must therefore be avoided by using a thorough training programme. It must always be remembered that the driving horse has the added burden of having to move within a gait, it cannot use its own natural rhythm.

The progression through the natural paces of the horse provides the greatest efficiency of energy utilisation, as measured by oxygen consumption. Each horse has its own speed within a pace at which it uses energy most efficiently. Horses working in harness actually have a greater energy demand because they are unable to move freely, so when selecting driving horses it is essential to choose those which inherently have the power, pace and stamina for the work. For a pair or team, it is of vital importance that they are complimentary in terms of temperament and that their gaits are fully compatible.

Training the Polo Pony

The game of polo was first described in a Persian manuscript dated 600 B.C., which means that the game has been played for over 2500 years. It is a very demanding sport, taxing ponies both physically and mentally. While each chukka only lasts seven minutes, the pony is on the go all of that seven minutes, with sudden changes of pace and direction.

After basic fitness has been achieved, schooling plays a very important part in the training programme; the pony is taught to be agile and able to adjust its feet quickly and nimbly during only slight changes of direction. To accomplish this, the pony is taught to be slightly behind the bit and off its forehand.

Basic Training

It is generally accepted that it takes two years to 'make' a polo pony. Ponies are usually taught the basics when they are four years old; as five-year-olds they play minor competitions and are ready for a full season of matches when they are six.

The basic training of a polo pony is no different from that of any other riding horse. The pony is taught to move freely forwards in walk and to trot in a balanced manner, to move away from the leg and canter on the correct lead.

Specialist Training

There are three main movements that need to be taught to the pony. These are:

- to change legs and alter balance immediately when asked
- to stop and start abruptly from any pace
- to turn on the haunches

The polo pony's ability to change legs at the canter at the slightest indication is vital if it is to be balanced at all times. The fast changes in pace and direction needed in the game of polo demand that the horse is always balanced.

A young, green pony will find sudden changes of pace very difficult and tiring. If these movements are introduced before the pony is fit, there is a risk of strains and sprains. Thus teaching a pony to stop abruptly from any pace begins by teaching it to stop instantly and correctly from walk. Once the pony understands the aids and gets fitter, faster paces can be introduced.

The turn on the haunches is again difficult for the youngster, and puts considerable strain on the hocks. If the movement is introduced at walk, the benefits are twofold; the pony learns the aids without being hurried and confused, and the hocks become gradually strengthened.

Once the pony has mastered these movements, it can be introduced to stick and ball. It is not advisable to hit a ball off a pony until its training is well underway and it is under full control. Time and patience are needed at this stage if the pony is not to be frightened. It is also wise not to over-corn the pony so that it is not too fresh.

There are several other things that may frighten a pony when it is starting to play polo. These include meeting other ponies coming towards it at speed, meeting swinging sticks, meeting the ball and being 'ridden-off' by another pony. These circumstances should all be encountered in training so that the pony gains confidence.

Fitness Programme

The polo season begins at the end of April and ends in September. Generally, a pony would play one or two 7 minute chukkas plus one practice session per week. (Low-goal matches consist of four chukkas, high-goal games of six.) A fit polo pony must thus be capable of galloping, twisting and turning for 14 minutes, two or three times a week. If this degree of work is to be maintained for the entire season of four months, the pony needs to be very well prepared and fit.

Two months is the minimum time in which it is possible to prepare a pony for match playing. If the pony is to stay fit throughout the season, it is wise not to overplay it for the first month of matches, ideally restricting it to one chukka per match. It is important to try to maintain optimum performance for as long as possible without overstressing the pony. Too much stress increases the incidence of injury and may also sour the pony. If the pony is to respond willingly to the demands of the game, it should not be placed in any situation that it is not capable of coping with.

Polo ponies are brought up at the end of February or the beginning of March. There is an initial period of walking work lasting 1 to 2 weeks. Throughout the training programme, one pony will be ridden with two being led from it. The ponies are ridden in rotation so that their skin becomes hardened and they are used to carrying the rider's weight. Trotting is usually introduced early on in the training programme and gradually built up, so that by the end of week 4 ponies are being worked for 90 minutes, 30 minutes of which would be trotting. This basic programme serves to get the horses muscled up, develops stamina and 'hardens' the legs. The polo pony also needs to develop its gymnastic and athletic ability, and schooling exercises should be introduced towards the end of the first month. These exercises would involve the movements previously described and some stick-and-ball work.

Canter-work begins 3 to 4 weeks before the first match is scheduled. Canter work is considered very important, with ponies being given up to 20 minutes' canter-work a day plus road work. Occasional fast work may be given prior to and during the season. This would consist of a short, half speed gallop to act as a 'pipe-opener' and to check that the pony is correct in its wind. Too much galloping is not recommended, as it tends to make ponies overexcited and gets them on their forehand.

Polo ponies are worked hard for short periods of time during a match, and the game demands explosive bursts of acceleration and great muscle power. Once basic fitness has been obtained through 4 to 6 weeks of slow work, training must concentrate on building up muscle power. These powerful, accelerating muscles are composed of a majority of fast twitch low-oxidative fibres which rely on anaerobic respiration. The fibres must be encouraged by performing in training the movements executed in the match, which means that schooling sessions are essential, not only in terms of obedience but also for fittening.

In scientific terms, polo ponies would be more efficiently trained by giving short intervals of maximal work, i.e. sprinting. This would bring about maximal development of the anaerobic system when combined with schooling exercises. However, the constraints of the labour force available mean that ponies are usually exercised in threes at a steady canter.

Conclusion

The theory and practice of getting horses fit is a vast subject, and no book can hope to do more than skim the surface. The aim of this book has been to make the owner/trainer/rider more aware of the systems of the horse's body and how these systems cope with exercise and adapt to training. A description has been given of various training methods, both traditional and modern, together with methods of assessing fitness. Using this new knowledge, trainers can select the method that suits them and their horse best. It must be emphasised that one method may suit one person but not another; horses, too, are individuals and must be treated accordingly. A degree of flexibility is vital in any training programme, and this is recognised to a greater extent in modern training methods than in traditional ones.

The trainer must be aware of the competitive goals and the standard of fitness necessary to achieve them. Overfitness and overfeeding can cause as many problems as lack of fitness, particularly for the inexperienced or younger rider.

Training horses cannot be learned out of a book, and there is no substitute for experience. However, it is the author's hope that this book will make gaining that experience a little easier for both horse and rider.

Appendix: A Guide to Speeds and Paces

Gait	Speed		
	km/h	m.p.m.	m.p.h.
Brisk walk	6	100	3.7
Slow trot	9	150	5.5
Medium trot	12	200	7.5
Fast trot	15	250	9.3
Slow canter	18	300	11.2
Medium canter	21	350	13.0
Brisk canter	24	400	14.8
Hand gallop	27	450	16.8
Medium gallop	30	500	18.7
Fast gallop	36	600	24.3

Glossary

Abdomen: body between chest and pelvis.

Acidosis: disturbance of acid-base balance due to any acid other than carbon dioxide.

Adenosine diphosphate: low energy compound.

Adenosine triphosphate: high energy compound involved in muscle contraction.

Adrenalin: substance secreted by medulla of adrenal gland. Causes rise in blood pressure, dilation of pupils, inhibited movement of the alimentary tract, sweating, fast breathing and the formation of glucose from liver glycogen.

Aerobic threshold: the point at which lactic acid builds up in tissue due to anaerobic respiration.

Albumin: serum protein.

Anaerobic: in the absence of oxygen.

Anterior mesenteric artery: supplies small intestine, caecum and part of colon.

Aorta: trunk of main artery beginning at base of left ventricle.

Ascarid: nematode parasite; whiteworm.

Aspergillus: genus of fungi, including several common moulds.

Atrial fibrillation: condition in which the heart beats irregularly due to rapid and ineffectual contractions of first chambers. Causes reduced performance.

Atrio-ventricular valves: those guarding opening between first and second chamber (atrium and ventricle) on left and right sides of heart (bicuspid and tricuspid valves).

Atrium (auricle): first chamber of the heart.

Azoturia: painful condition of large muscle masses of back and behind quarter. Also known as setfast; tying-up.

Bars: continuation of wall of foot; turns inward at heel to run parallel with frog.

Basophil: white blood cells identified by dye-staining at post mortem.

Biochemistry: the science of organic reactions taking place within the body.

Biological value: the amino-acid make up of the protein related to its nutritional value.

Biscuspid valve: valve separating left atrium and left ventricle of heart.

Blood: fluid circulating in arteries, capillaries and veins and driven by pumping action of the heart.

Blood test: examination of blood samples in laboratory to aid diagnosis of disease.

Bone: hard substance making up body's skeleton.

Bronchiolitis: inflamed bronchioles when tubes become full of exudate.

Bursa: sac or cavity filled with fluid at places where friction is likely to occur.

Caecum: large comma-shaped sac between small intestine and colon.

Calcify: to become impregnated with calcium, i.e. bony.

Calcium: mineral required for bone and tooth formation.

Cancellous: spongy (used in connection with bone tissues).

Carbohydrate: food substance found in vegetable and animal tissue.

Carbon dioxide: a gaseous waste product of aerobic respiration.

Carbon fibre implant: a treatment for damaged tendons of the lower limb.

Cardiac muscle: muscle of the heart.

Cellulose: carbohydrate making up part of plant cell wall.

Circulatory system: blood, blood vessels, heart, lymph and lymphatic vessels.

Compact bone: dense outer layer of bone.

Corium: modified vascular tissue inside horn or hoof. Can be likened to bed of human nail.

Coronet band: band at the top of hoof and lower part of pastern; seat of hoof growth.

Creatine phosphate: a source of energy contained in limited amounts in muscle cells.

Deep digital flexor tendon: tendon of the lower leg attaching to the pedal bone.

Dental system: the teeth and associated structures.

Detoxify: to render poisons harmless.

Digestible energy: a measure of the energy value of feedstuffs, measured in megajoules.

Digestive system: the organs and glands involved in the digestion of food.

Digital cushion: fibro-elastic fatty pad at back of foot.

Egg count: a count of parasitic worm eggs in the faeces.

Elastic zone: range of stress that bone can accommodate by stretching.

Electrocardiogram: a graphic tracing of the electric current produced by the heart muscles.

Electrolyte: substance present in the body fluids which is capable of conducting electricity in various body functions, such as nerve impulses, oxygen, and carbon dioxide transport and muscle contraction.

Ephysema: an abnormal accumulation of air in tissues or organs, e.g. overinflation of the alveoli of the lungs.

Endotoxin: a poison produced within the body.

Enotenon: internal tendon covering supplying blood vessels and nerves.

Enzyme: a protein which catalyses (helps cause or accelerate) a chemical reaction without being consumed in the process.

Eosinophil: type of white blood cell.

Epiglottis: small flap of cartilage which covers entrance to voice box.

Epiphyseal: 'growth' plate; the cartilage separating the epiphysis from the shaft of a bone.

Epiphysis: a part of a bone, especially at the end of a long bone, which develops separately from the shaft of the bone during the growth period. During this time it is separated from the main portion of the bone by cartilage.

Epistaxis: nosebleed.

Epitenon: external covering of the tendon.

Erythrocyte: red blood cell which transports oxygen.

Excretion: the act of eliminating the body's waste materials.

Exhalation: as expiration.

Exhaustion: complete inability to continue work.

Exostosis: a benign bony growth projecting outward from the surface of a bone.

Expiration: the act of expelling air from the lungs.

Extracellular fluid: body fluid contained between the cells and bathing them.

Fartlek: a training method.

Fat: white or yellowish material laid down around, or in, various organs and muscles of the body. Also part of diet comprised of fatty acids and glycerol, high in feeds such as linseed.

Fatigue: tiring of the body systems associated with exercise.

Fermentation: enzymatic decomposition.

Fibre: insoluble carbohydrate, such as cellulose, making up an essential part of the diet.

Fibrillation: rapid contraction, as in atrial fibrillation.

Fibrin: the essential fibrous protein portion of the blood.

Fibrinogen: a protein in the blood essential to the clotting process.

Fibroblast: a specialised body cell.

Fibrous adhesion: a fibrous band or structure by which parts abnormally adhere.

Firing: applying a heated firing iron to a leg to produce a severe inflammation; used to treat chronic or subacute inflammations of the joints, tendons and bones.

Floating: the act of filing down the teeth to remove sharp edges.

Foramina: natural openings or passages in the body.

Forceplate: a plate used for recording the forces on a horses foot as it moves.

Founder stance: the typical stance of a horse with laminitis in which the animal will try to take as much weight as possible off its front feet.

Frog: the band of horny substance in the middle of the sole of a horse's foot, dividing into two branches and running towards the heal in the shape of a 'v'.

Gad-fly: the warble-fly.

Gait analysis: filming and analysing each portion of the horse's gait.

Gaseous exchange: the exchange of oxygen from the lungs for carbon dioxide in the blood.

Gastric: pertaining to the stomach.

Germ: a pathogenic organism.

Glands: aggregations of cells, specialised to secrete and excrete materials.

Globulins: blood proteins.

Glucose: a sugar which is a principal source of energy.

Glycogen: the body's main carbohydrate-storing substance.

Glycogen sparing: when the body uses fat as an initial source of energy, conserving glycogen stores.

Glycolysis: the breakdown of glycogen to provide energy within the body.

Haematoma: a localised collection of blood, usually clotted, in an organ, space or tissue, due to a break in the wall of a blood vessel.

Haemoglobin: the oxygen-carrying protein pigment of the red blood cells.

Haemorrhage: bleeding.

Haversian systems: the systems into which the bone cells are organised.

Heart: muscular organ with four chambers which pumps blood through system of vessels.

Heart rate monitor: a device for measuring the heart rate of a working horse.

Heat exhaustion: hyperthermia; circulatory collapse and shock caused by high environmental temperature, high humidity and poor ventilation.

Heat stroke: more serious than heat exhaustion; sweating usually stops. Often fatal.

Heaves: a respiratory ailment; forced expiration resulting from rupture of the alveoli in the lungs. Caused by allergies and dust.

Histamine: a chemical compound which dilates capillaries and constricts the smooth muscle of the lungs.

Hoof: horny casing of foot.

Hormone: a chemical substance, produced in the body by an organ, which regulates the activity of another specific organ.

Hyperthermia: heat exhaustion or heat stroke.

Hypertrophy: the enlargement or overgrowth of an organ.

Ileum: the distal portion of the small intestine extending from the jejunum to the caecum.

Impaction: the condition of being firmly lodged or wedged.

Incisor: front tooth specialised for biting.

Inflammation: a condition where tissues show pain, heat, redness, swelling and exudate as a reaction to injury.

Influenza: an acute viral infection involving the respiratory tract.

Ingestion: eating.

Inhalation: drawing air into the lungs.

Inspiration: as inhalation.

Intercostal: between ribs.

Interval training: a method of getting horses fit; a specific period of work is given, followed by a specific period of rest.

Intestinal: pertaining to intestines.

Intestinal flora: the bacteria normally present within the intestines.

Intracellular fluid: fluid found inside cells.

Jejunum: the middle portion of the small intestine, extending from the duodenum to the ileum.

Lactic acid: an organic acid normally present in muscle tissue. Produced by anaerobic muscle metabolism.

Laminae: membrane or sheet containing fine leaf-like projections.

Laminitis: inflammation of the laminae of the foot.

Larva: an early developmental stage of a worm.

Larynx: the voice box.

Lesion: an abnormal change in the structure of a part due to injury or disease.

Letting down: reducing a horse's fitness in a gradual and controlled manner.

Leucocytes: white blood cells or corpuscles.

Ligament: band of fibrous tissue that connects bones or cartilages.

Lignin: an indigestible structural carbohydrate found in plant cell walls.

Lipid: an organic substance containing fat, which is an important component of living cells.

Lungworm: nematode parasite living in the air passages of the lungs.

Lymph: a transparent yellowish liquid containing white blood cells and derived from tissue fluids.

Lymphangitis: inflamed lymphatic vessels and lymph nodes, especially in legs.

Lymphatic system: a system of vessels containing and transporting lymph.

Lymphocyte: a type of white blood cell.

Lysine: an essential amino acid.

Maintenance: a state where the animal is fed enough to maintain life and all body functions without gaining or losing weight.

Marrow: red or yellow soft material in central cavity of long bones.

Megajoule: unit in which the energy value of a foodstuff is measured; this is expressed as megajoules of digestible energy per kilogram (MJDE/kg).

Membrane: a thin layer of tissue which covers a surface, lines a cavity or divides a space or organ.

Mesentry: a fold or membrane attaching various organs to the body wall.

Metabolism: the total physical and chemical activities of an animal.

Methionine: an essential amino acid.

Micronutrients: nutrients needed in minute amnounts, e.g. minerals and vitamins.

Micro-organisms: minute microscopic organisms such as bacteria, viruses, moulds, yeasts and protozoa.

Micropolyspora faeni: a fungus whose spores can cause an allergic reaction.

Mineral: inorganic substances some of which are essential for the maintenance of health, e.g. iron, iodine, calcium.

Mitochondria: the powerhouses of the cell where energy is produced.

Molar: tooth adapted for grinding.

Monday morning disease: azoturia; painful movement and 'tying-up' of back muscles.

Monocytes: white blood cells active in fighting subacute infections.

Monosaccharide: the most simple form of sugar, e.g. glucose, fructose.

Mucus: the free slime of a mucous membrane; composed of the secretions of glands.

Murmur: a periodic sound, of short duration, of cardiac or vascular origin.

Muscle biopsy: the removal of a sample from living muscle using a specialised needle.

Mutant: a permanent change in the characteristics of an organism.

Myoglobin: a protein contributing to the colour of muscle and acting as a store of oxygen.

Nasal passages: the airways of the head, leading from the nostrils to the trachea.

Navicular bone: a small bone in the foot of a horse.

Nematode: any of a group of roundworms which are mainly internal parasites.

Nerves: cordlike structures, visible to the naked eye, comprising a collection of nerve fibres which convey impulses between a part of the central nervous system and some other region of the body.

Nervous system: the system of nerves throughout the body.

Neutrophil: a white blood cell type.

Nostril: entrance to the respiratory tract.

Organ: a somewhat independent part of the body that performs a special function or functions.

Osmosis: diffusion through a semipermeable membrane.

Osselet: a bony growth on the inner aspect of a horse's knee or on the lateral aspect of the fetlock.

Ossify: to change or develop into bone.

Osteitis: inflammation of a bone.

Osteoblasts: cells involved with bone destruction.

Osteoclosts: cells involved with bone formation.

Oxidative: a reaction using oxygen.

Oxygen: colourless, odourless gas inhaled from the atmosphere.

Oxygen debt: when, at maximal exercise levels, oxygen is used in order to stablise the body after exercise is finished.

Pacemaker: the specific area of the heart that controls the heartbeat.

Packed cell volume: proportion of blood cells to plasma; expressed as a percentage.

Parasite: a plant or animal which lives upon or within another living organism at whose expense it obtains some advantage.

Pedal bone: bone inside hoof.

Periosteitis: inflammation of the periosteum (bone covering).

Periople: external, shiny, water-repellant covering of the hoof.

Phagocyte: any cell that ingests micro-organisms or other cells and foreign particles.

Pharynx: the sac between the mouth and the oesophagus.

Physiology: the branch of biology which deals with the normal functions and activities of life or living matter.

Pinch test: a simple test to assess dehydration by pinching a fold of skin on the neck.

Plasma: the liquid portion of the blood.

Poultice: a moist dressing applied hot to a given area in order to create moist local heat or counter-irritation.

Protein: one of a group of complex compounds which contain nitrogen and are composed of amino acids.

Protozoa: single-celled organisms.

Pulmonary: pertaining to the lungs.

Pulse: rhythmic throbbing of an artery which may be felt by the finger. Caused by blood forced through the vessel by contractions of the heart.

Pus: a liquid inflammation product made up of white blood cells and dead tissues.

Quarters: part of the wall of the foot between the heel and the toe.

Quidding: when the horse spits out partially chewed food.

Rasping: filing the teeth with a rasp to provide dental care.

Ration: a carefully designed allocation of food.

Recovery rate: the time in which the horse recovers from exertion; measured by pulse and respiration rates.

Red blood cells: haemoglobin-carrying cells in the blood.

Remodelling: the process undergone by bone which has been put under stress.

Renal artery: supplies the kidneys with blood.

Reproductive system: the organs involved with conception, gestation and birth.

Respiratory rate: number of breaths in one minute.

Respiratory system: the airways and lungs.

Rhinopneumonitis: inflammation of the nasal and pulmonary mucous membrane (equine herpes virus 1).

Rickets: disease of young horses characterised by lack of calcium in bones.

Ring bone: bony enlargements below fetlock.

Roughing-off: gradually adapting the horse to living outside rather than being stabled.

Rubefacient: drug or other substance that increases blood flow.

Scouring: diarrhoea; loose, runny faeces.

Secrete: to produce and give off cell products.

Semilunar valves: half-moon shaped cups forming a valve which guards the exits of the heart.

Serum: clear fluid which separates red blood clot, i.e. whole blood minus cells and fibrinogen.

Sesamoid bones: small bones inserted into tendons where pressure occurs.

Setfast: see azoturia.

Shaft: main part of long bone.

Skeletal muscle: striated or voluntary muscle responsible for movement of the skeleton.

Skeletal system: the body's framework of bone and cartilage.

Small airway obstruction: blockage of bronchioles of lungs due to excess mucus production and muscle contraction.

Smooth muscle: not under voluntary control; found in walls of bladder and uterus.

Soft palate: membrane of muscles separating mouth from pharynx.

Sole: undersurface of foot.

Sore shins: inflamed lining of cannon bone.

Spavin: condition of hock joint or surrounding area.

Spleen: a blood storage organ situated near the stomach.

Splint: condition of the splint bone.

Spongy bone: soft bone.

Starch: the form in which plants store glucose.

Stethoscope: instrument for listening to heart and gut sounds.

Stress: forcible exerted influence or pressure.

Strongyle: redworm.

Subclinical: where the horse is suffering from disease but shows no clinical symptoms.

Sugar: carbohydrate found in vegetable and animal tissue, e.g. starch, sucrose, lactose.

Sweat: salty fluid; evaporation from skin causes cooling of the body.

Synchronous diaphragmatic flutter: a condition where the heart appears to be beating in the flanks, indicating severe fatigue. Also known as thumps.

Synovial fluid: fluid found within joint; has a lubricating function.

Systemic circulation: the circulation of blood through the body, excluding the lungs.

Systole: contraction of the heart chambers.

Temperature: 38°C (100.5°F) in a healthy horse.

Tendon: a fibrous cord of connective tissue which attaches muscle to bone.

Tendon sheath: a productive connective tissue layer surrounding the tendon.

Tendon splitting: a method of treating damaged tendons.

Tenoblasts: cells involved in tendon destruction.

Tenocytes: cells involved in tendon formation.

Thermography: a method of recording the heat of body surfaces.

Thumps: see synchronous diaphragmatic flutter.

Tissue: an aggregation of similarly specialised cells united in the performance of a particular function.

Tissue respiration: the production of energy in the cells.

Toe: front of hoof.

Trachea: the windpipe.

Tricuspid valve: the valve lying between the right atrium and right ventricle in the heart.

Triglyceride: a type of fat.

Tropocollagen: the precursor of collagen.

Tushes: canine teeth found in male but not usually female.

Tying-up: see azoturia. May also be result of a pulled muscle.

Ultrasound: controlled doses of high frequency sound (radiation) used for therapeutic treatment.

Urea: waste product discharged in the urine. Contains nitrogen.

Urinary system: the kidneys, ureter, bladder and uretha.

Vaccine: a suspension of attenuated or killed micro-organisms administered for the prevention or treatment of infectious disease.

Vasoactive: causes blood vessels to constrict or dilate.

Vein: a vessel through which the blood passes from various organs back to the heart.

Vena cava: the major vein of the body.

Ventricles: the two larger, muscular chambers of the heart.

Villi: tiny finger-like extensions of a membrane.

Virus: one of a group of minute infectious agents.

Vitamin: organic substance necessary for normal metabolism.

Vocal cords: membranes, contained within the pharnyx, which allow vocalisation.

Warm-up: a period of exercise prior to work to prime the body systems.

White line: junction of wall of foot with sole.

Wolf teeth: vestigial pre-molar which may erupt and have to be removed.

Index